RARE EARTHS AS A PROBE OF ENVIRONMENT AND ELECTRON-PHONON INTERACTION IN OPTICAL MATERIALS

Rare Earths as a Probe of Environment and Electron-Phonon Interaction in Optical Materials

Rosanna Capelleti
Andrea Baraldi
Elisa Buffagni
Nicola Magnani
and
Margherita Mazzera

Nova Science Publishers, Inc.
New York

Copyright © 2009 by Nova Science Publishers, Inc.

All rights reserved. No part of this book may be reproduced, stored in a retrieval system or transmitted in any form or by any means: electronic, electrostatic, magnetic, tape, mechanical photocopying, recording or otherwise without the written permission of the Publisher.

For permission to use material from this book please contact us:
Telephone 631-231-7269; Fax 631-231-8175
Web Site: http://www.novapublishers.com

NOTICE TO THE READER

The Publisher has taken reasonable care in the preparation of this book, but makes no expressed or implied warranty of any kind and assumes no responsibility for any errors or omissions. No liability is assumed for incidental or consequential damages in connection with or arising out of information contained in this book. The Publisher shall not be liable for any special, consequential, or exemplary damages resulting, in whole or in part, from the readers' use of, or reliance upon, this material.

Independent verification should be sought for any data, advice or recommendations contained in this book. In addition, no responsibility is assumed by the publisher for any injury and/or damage to persons or property arising from any methods, products, instructions, ideas or otherwise contained in this publication.

This publication is designed to provide accurate and authoritative information with regard to the subject matter covered herein. It is sold with the clear understanding that the Publisher is not engaged in rendering legal or any other professional services. If legal or any other expert assistance is required, the services of a competent person should be sought. FROM A DECLARATION OF PARTICIPANTS JOINTLY ADOPTED BY A COMMITTEE OF THE AMERICAN BAR ASSOCIATION AND A COMMITTEE OF PUBLISHERS.

LIBRARY OF CONGRESS CATALOGING-IN-PUBLICATION DATA

ISBN: 978-1-60692-137-1

Available upon request

Published by Nova Science Publishers, Inc. New York

Contents

Preface — vii

Chapter I   Introduction — 1

Chapter II   Electronic Energy Level Scheme of Rare Earths — 7

Chapter III RE^{3+} as a Probe of the Environment — 35

Chapter IV   Electron-Phonon (e-p) Interaction — 53

Chapter V   Conclusion — 65

Acknowledgments — 67

References — 69

Index — 74

PREFACE

The rare earth electronic transitions, widely exploited for optical applications, occur within the shielded 4f shell and originate very sharp absorption (or emission) lines in spectra of crystals measured at low temperatures. High resolution absorption spectroscopy, covering broad spectral and temperature ranges, is a powerful tool not only to supply a complete and precise energy level scheme for the rare earth in optical materials, but also to disclose finer details on rare earth environment and interactions with i) other rare earths, ii) lattice vibrations, and iii) nuclear magnetic moments. In the present work the experimental results, taken in the temperature range 9-300 K, covered a spectral region as wide as 75-25000 cm^{-1} at a resolution as fine as 0.01 cm^{-1}. Their interpretation is supported by calculations based mainly on crystal field theory, superposition model, and nuclear-electronic hyperfine interaction analysis. In this framework, a review is presented on the results obtained by our group during the last years in single crystals (BaY_2F_8 and $YAl_3(BO_3)_4$ doped with a variety of rare earths for laser and self frequency doubling applications), in Er-doped glasses, and in nanostructured glass-ceramics for integrated optics.

Chapter I

INTRODUCTION

Rare earths (RE) doped into transparent host crystals, thanks to the 4f electron localization, make it possible to investigate fundamental physical phenomena with incomparable accuracy and precision and continuously open new fields to optical technology. Only few applications may be mentioned, as long-lived laser sources, self frequency doubling devices, fast and efficient scintillators, erbium doped fiber amplifier, and new materials for optical data storage and processing [1-5].

According to the Chemistry textbooks the RE 4f intraconfigurational transitions occur within the electrostatic shield provided by the outermost 5s and 5p shells. As a consequence, the related optical spectra maintain much of an atomic-like character and should be affected only to a very limited extent by the RE ion surrounding. As a matter of fact tables, supplying the RE^{3+} levels diagram in a specific host (e.g. $LaCl_3$ [1,6]), are of a great use for a preliminary identification of the lines due to a given RE^{3+} within the optical spectrum of a new material. This common trend is exemplified for one of the most investigated RE^{3+}, as Er^{3+}, in Figure 1, where the room temperature (RT) absorption spectra are displayed in the region of the $^4I_{15/2} \rightarrow {}^4I_{13/2}$ transition of Er^{3+} hosted in different crystalline, polycrystalline, amorphous matrices, and nanostructured materials. In the spectra of crystalline and polycrystalline materials, some structure can be envisaged, however the overlapping among the absorption lines is remarkable, precluding any correct line identification and attribution. On the contrary, if the optical measurements are performed at low temperatures (e.g. 9 K) and at high resolution (where necessary), the spectra reveal more specific features of a given host matrix, see Figure 2. Thus the RE^{3+} may be regarded as a very sensitive probe of the surrounding, in spite of the shield supplied by the outermost 5s and 5p shells.

An inspection of the 9 K spectra discloses some finer details on Er^{3+} environment and interactions.

1) The lines, in crystalline and polycrystalline matrices, narrow by lowering the temperature (compare curves a-f in Figure 1 with curves a-f in Figure 2, respectively), therefore a detailed study of the line width (and position) vs. temperature can be exploited to study the electron-phonon interaction, see Secs. IV A and IV B.
2) The lines are very sharp in slightly doped samples, but they broaden by increasing the Er^{3+} concentration, compare the two curves in Figure 3, which are a magnification of curves a and b of Figure 2 in the range 6550-6595 cm^{-1}. In amorphous matrices, like silica, the spectra remain broad and almost structureless even at low temperature (compare curves i in Figures 1 and 2), while in nanostructured glassceramic the lines sharpen if Er^{3+} is embedded in SnO_2 nanocrystals (compare curves h in Figures 1 and 2), therefore the RE^{3+} line width behaves as a probe of the long range order within the matrix, see Sec. III A 1.
3) Additional weaker lines appear in heavily Er^{3+} doped BaY_2F_8 (20 at% Er^{3+}), which are absent in slightly doped sample (0.5 at% Er^{3+}), compare curves a and b in Figure 3, and monitor weak short range Er^{3+}-Er^{3+} interaction caused by the presence of loosely bound Er^{3+} clusters, see Sec. III B.
4) The line positions, separations, and number work as a signature of Er^{3+} in a given crystalline or polycrystalline host. In the same matrix (e.g. BaY_2F_8) the main line spectra are essentially the same, except for the line amplitudes, compare curves a and b in Figure 2 for two different Er^{3+} concentrations. This result is a consequence of the peculiar crystal field probed by the Er^{3+} in each matrix, see Sec. II C.

The number of lines is rather large, see for example Figure 2, where only one (i.e. $^4I_{15/2} \rightarrow {}^4I_{13/2}$) among the many Er^{3+} intra-configurational transitions is considered. Therefore it is necessary to follow joined experimental and theoretical approaches to assign correctly each line to a specific transition from the sublevels of the ground manifold to those of different excited manifolds. The aim is to provide a complete RE^{3+} level scheme, of interest both from basic and applicative point of view: the task can be successfully accomplished if high resolution spectra are available for the largest number of transitions, i.e. over the widest spectral range (Sec. II).

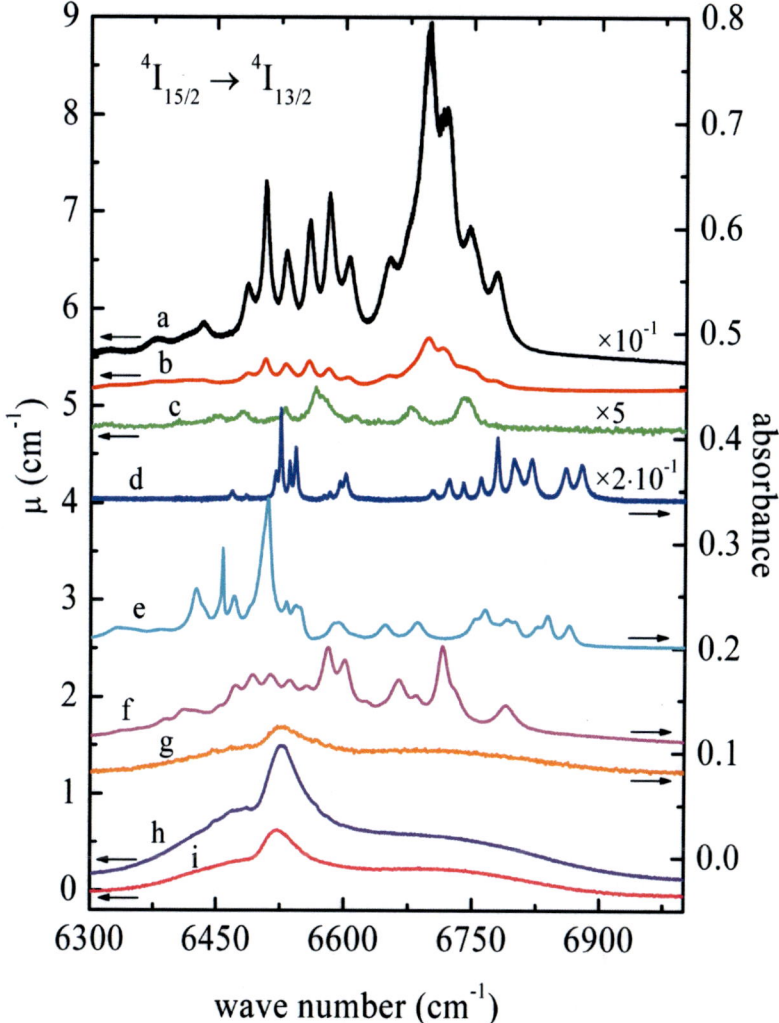

Figure 1. Optical absorption spectra measured at room temperature in the region of Er^{3+} $^4I_{15/2} \rightarrow {}^4I_{13/2}$ transition for different insulating materials. Curve a: BaY_2F_8: Er^{3+} 20 at% single crystal (the amplitude is reduced by a factor 10); curve b: BaY_2F_8: Er^{3+} 0.5 at% single crystal; curve c: YAB: Er^{3+} 0.3 mol% single crystal (the amplitude is multiplied by a factor 5); curve d: YAG: Er^{3+} 2 mol% single crystal (the amplitude is divided by a factor 5); curve e: Er_2O_3 polycrystalline powders (pellet); curve f: ErF_3 polycrystalline powders (pellet); curve g: SnO_2: Er^{3+} 2 mol% polycrystalline powders (pellet); curve g: SiO_2: SnO_2 16 mol%, Er^{3+} 1 mol% glass-ceramic; curve i: SiO_2: Er^{3+} 0.5 mol% glass. The ordinate scale for each curve is indicated by an arrow: the absorbance is used if the sample thickness cannot be evaluated correctly (e.g. pellets).

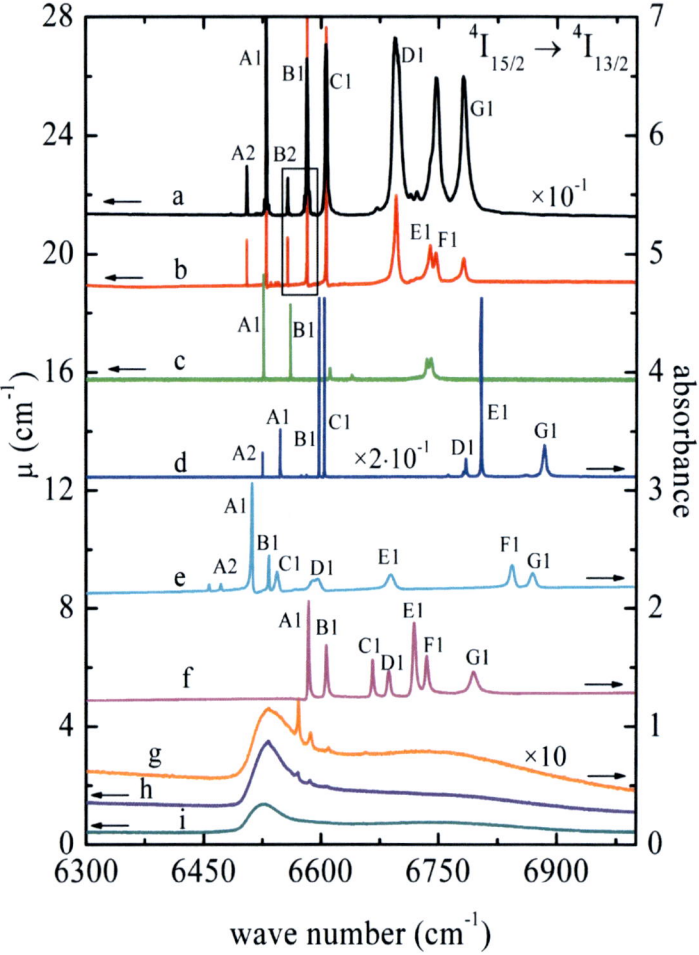

Figure 2. Optical absorption spectra measured at 9 K in the region of Er^{3+} $^4I_{15/2} \rightarrow {}^4I_{13/2}$ transition for different insulating materials. Curve a: BaY_2F_8: Er^{3+} 20 at% single crystal (the amplitude is reduced by a factor 10), res.=0.4 cm^{-1}; curve b: BaY_2F_8: Er^{3+} 0.5 at% single crystal, res.=0.02 cm^{-1}; curve c: YAB: Er^{3+} 0.3 mol% single crystal, res.=0.04 cm^{-1}; curve d: YAG: Er^{3+} 2 mol% single crystal (the amplitude is divided by a factor 5), res.=0.4 cm^{-1}; curve e: Er_2O_3 polycrystalline powders (pellet), res.=0.5 cm^{-1}; curve f: ErF_3 polycrystalline powders (pellet), res.=0.5 cm^{-1}; curve g: SnO_2: Er^{3+} 2 mol% polycrystalline powders (pellet) (the amplitude is multiplied by a factor 10), res.=0.5 cm^{-1}; curve g: SiO_2: SnO_2 16 mol%, Er^{3+} 1 mol% glass-ceramic, res.=0.1 cm^{-1}; curve i: SiO_2: Er^{3+} 0.5 mol% glass, res.=0.5 cm^{-1}. The ordinate scale for each curve is indicated by an arrow: the absorbance is used if the sample thickness cannot be evaluated correctly (e.g. pellets).

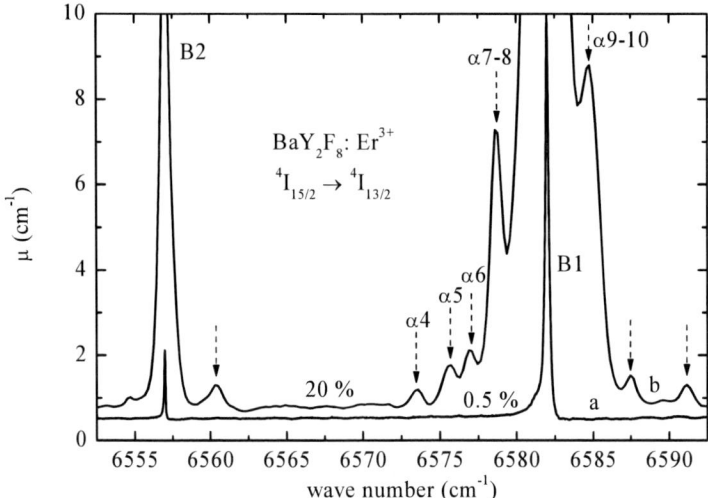

Figure 3. Magnification (in the wave number range indicated by the rectangle drawn in Figure 2) of the optical absorption spectra measured at 9 K in the region of Er^{3+} $^4I_{15/2} \rightarrow {}^4I_{13/2}$ transition for BaY_2F_8 single crystals doped with different Er^{3+} concentrations. Curve a: 0.5 at%; curve b: 20 at%. The dotted arrows put in evidence the weak additional lines.

On the basis of the above considerations, the spectra were measured in the temperature range 9-300 K and covered the spectral region 75-25000 cm^{-1} at a resolution as fine as 0.01 cm^{-1}. Their interpretation was supported by calculations based mainly on crystal field theory and superposition model. The materials investigated were single crystals, as BaY_2F_8 (BaYF) and $YAl_3(BO_3)_4$ (YAB), doped with a variety of RE^{3+} (Ce^{3+}, Dy^{3+}, Ho^{3+}, Er^{3+}, Tm^{3+}, and Yb^{3+}) for laser and self frequency doubling applications, Er^{3+}-doped nanostructured glassceramics for integrated optics, and polycrystalline powders of ErF_3 and Er_2O_3, where Er^{3+} is a member of the host matrix, see Sec. II A.

In Sec. II a comprehensive analysis of the RE^{3+} spectra as a function of temperature and light polarization will be accompanied by a critical discussion i) of the crystal field parameters along the lanthanide series and ii) of the wavefunction symmetries.

In Sec. III subtle effects are evidenced by applying the high resolution spectroscopy at low temperature, e.g. i) to reveal the weak interactions between well shielded rare earths to originate loosely bound clusters in crystals, ii) to distinguish the rare earths localized within crystalline nanoclusters from those dispersed in the amorphous silica matrix in glass-ceramics, and iii) to detect a variety of Ho^{3+} hyperfine structure patterns in YAB over a wide spectral range.

In Sec. IV the electron-phonon (e-p) interaction is considered. The RE^{3+} spectral line-position, -shape, and -width are analyzed as a function of temperature and by exploiting current models for line-broadening and shift: the aim is to get the e-p coupling constants and the energy of the coupled phonons. The vibronic replica of the whole zero-phonon line series by well identified lattice phonons are detected as a consequence of the simultaneous excitation of an electronic and a vibrational transition.

Chapter II

ELECTRONIC ENERGY LEVEL SCHEME OF RARE EARTHS

A. DETAILS ON THE SAMPLES, SAMPLE PREPARATION, AND FT MEASUREMENTS

Different kinds of samples were investigated: 1) single crystals as BaY_2F_8 and $YAl_3(BO_3)_4$; 2) polycrystalline powders and films, as ErF_3, Er_2O_3, and SnO_2; 3) glasses as silica; and 4) nanostructured glassceramics. BaYF single crystals were grown by a computer-controlled Czochralski technique under a purified argon atmosphere at Physics Department of the University of Pisa (Pisa, Italy) [7]. The samples were prepared from 5N pure BaF_2, Y_2F_3, and RE_2F_3.

YAB single crystals were grown by means of the top-seeded solution growth (TSSG) method from a $K_2O/MoO_3/B_2O_3$ flux at Research Institute for Solid State Physics and Optics (HAS, Budapest, Hungary). RE^{3+} was added as RE oxide (RE_2O_3). Detailed description of the experimental conditions can be found in Ref. [8].

In both crystals the RE^{3+} substitution for the homovalent Y^{3+} does not require any charge compensation, thus rather high RE^{3+} solubility can be obtained: the RE^{3+} segregation coefficient was close to unit [9-11]. For optical measurements the samples were x-ray oriented, cut, and polished.

Glassceramics of Er-doped silica containing SnO_2 nanoclusters were prepared by the sol-gel technique at Department of Materials Science of the University of Milano-Bicocca (Milano, Italy). Tetraethoxysilane (TEOS), dibutyl tin diacetate, and erbium nitrate were co-gelled: after gelation and drying the samples were heated in oxygen up to 1050 °C to induce SnO_2 nanoclustering and silica densification [12,13]. Fluorinated samples were also synthesized by co-hydrolyzing TEOS and triethoxyfluorosilane [14]. The sol-gel route was also employed to prepare reference samples of Er doped silica and microcrystalline powders of SnO_2: Er [15]. The final bulk samples were optical grade disks of approximately 2 cm diameter and 1 mm thickness.

Table 1. Materials, space groups, preparation methods, dopants, and concentrations considered

Material	Space group	Preparation method	Dopant	$C_\%$	d (mm) $Wt_\%$
BaY_2F_8 single crystals (BaYF)	C2/m	computer-controlled Czochralski [7]	-	-	3.2
			Er	0.5	2.80
				2	1.96
				12	3.6
				20	0.9/7.53
			Dy	4.4	2.8
			Tm	0.5	2.56
				5	3.09
			Tm, Ho	5.2, 0.5	2.38
			Ce	0.24	0.84
				3.75	1.72
$YAl_3(BO_3)_4$ single crystals (YAB)	R32	top-seeded solution growth [8]	Er	0.3	0.58
				0.5	1.5
				12	1.9
			Dy	1	1.98
			Ho	1	1.99
Er_2O_3 polycrystalline powders	Ia3	purchased	-	-	*9.5*
ErF_3 polycrystalline powders	Pnma	purchased	-	-	*12.1*
SnO_2 polycrystalline powders	$P4_2/mnm$	sol-gel [15]	Er	2	*58.3*
SiO_2 glasses		sol-gel [14,16]	Er	0.5	*0.57*
			Er, F	0.2, 2	*1.33*
				0.2, 5	*1.5*
SiO_2: SnO_2 glass ceramics		sol-gel [12,13,15,27]	Er, SnO_2	1, 8	*1.18*
				1, 16	*1.22*

The RE doping level $C_\%$ (last but one column) is given in terms of molar fraction (mol%), except for BaY_2F_8 samples where the concentration is expressed in atomic fraction (at%). The last column reports the thickness d (expressed in mm) for bulk samples and the weight ratio $Wt_\%$ with respect to CsI or KBr for pellets (in italic).

At variance with BaYF and YAB, in glasses and glassceramics the trivalent Er substitutes for tetravalent silicon (or tin): charge compensation is required and may result from oxygen substitution by either OH^- or F^- (the latter in fluorinated glasses).

Samples with different RE concentrations and thicknesses were investigated, see Table 1. The RE doping level is given in terms of atomic fraction (at%), i.e. the ratio of N_{RE} to N_Y for BaY_2F_8 crystals, where N_X is the number of X ions. In the case of YAB crystals, sol-gel glasses, glassceramics, and microcrystalline powders the dopant concentration (RE, Sn, F) is expressed in terms of molar fraction (mol%).

To measure the infrared (IR) spectra in the high absorption region, pellets were prepared from chips of BaYF and YAB single crystals, glassceramics, and microcrystalline powders (Er_2O_3, ErF_3, and SnO_2: Er). They were finely ground and mixed with KBr or CsI powders, which are transparent in the spectral range of interest. The weight ratio of the material under investigation to that of KBr or CsI is reported in Table 1.

The spectra were measured by means of a Fourier transform spectrometer Bomem DA8, operating in vacuum and in a wide wave number range (75-25000 cm^{-1}). Such a large spectral extension was made possible by using different radiation sources (high pressure Hg vapour lamp, SiC globar, and quartz halogen lamp), beam-splitters (3, 6, and 12 μm mylar, Ge-covered KBr, Sb_2S_3-covered CaF_2, and TiO_2-covered quartz), and detectors (DTGS, wide range liquid nitrogen cooled MCT, liquid nitrogen cooled cold filter InSb, and Si photodiode). The apodized resolution was as fine as 0.02 cm^{-1}: this value was improved to 0.01 cm^{-1} by applying a boxcar apodization.

Measurements with linearly polarized light in the IR region (2000-14000 cm^{-1}) were performed by using a gold grid polarizer deposited onto a KRS5 substrate: a home made polarizer holder allowed changing the direction of the light electric field without breaking the vacuum in the spectrophotometer sample compartment.

The sample temperature was varied between 9 and 300 K by assembling the sample in a 21SC Model Cryodine Cryocooler of CTI Cryogenics equipped with polyethylene, KRS5, and quartz windows, for measurements in far IR, middle and near IR, and visible range, respectively.

B. ABSORPTION SPECTRA ANALYSIS, LINE ATTRIBUTION ON THE BASIS OF THE EXPERIMENTAL DATA

Because of the electrostatic and spin-orbit interaction, the $4f^n$ configuration (with n=1, 2, 3, ..., 13) of the free RE^{3+} ion is split into a number of levels or manifolds which may be as high as 327 for the Gd^{3+} in the middle of the lanthanide series, whilst for the end members (Ce^{3+} and Yb^{3+}) the number is reduced to two [6]. Each manifold is labeled as $^{2S+1}L_J$, according to the values of total spin (S), total orbital momentum (L), and total angular momentum (J), see Sec. II C. The energy region covered by the levels is very wide, spanning from IR to ultraviolet (uv), therefore only spectrometers covering a very broad wave number range (Sec. II A) are able to provide a full description of the energy levels in a given material and to supply copious data bases for a sound theoretical interpretation, see Sec. II C. For the RE^{3+} end members (Ce^{3+} and Yb^{3+}) the transitions are limited to IR, while in the case of Gd^{3+} even the lowest energy transition falls in the uv [1].

A further level splitting is induced by the crystal field, if the ion is embedded into a crystal lattice. The number of sublevels depends on the crystal field symmetry and on J, up to a maximum of (2J+1)/2 for ions with an odd number of 4f electrons (Kramers doublets) and up to 2J+1 for ions with an even number of 4f electrons, in the case of low symmetry crystal fields (e.g. monoclinic), see Table 1 and Sec. II C. By assuming as an example Er^{3+} (with a $4f^{11}$ configuration and 41 predicted manifolds [6]), the lowest two manifolds $^4I_{15/2}$ and $^4I_{13/2}$ are split into 8 and 7 sublevels, respectively, in monoclinic BaYF [10], in trigonal YAB [11], in Er_2O_3 (space group Ia3), and in ErF_3 (space group Pnma) [17], see Figures 10-13. This explains why the related 9 K absorption spectra displayed in the region of the $^4I_{15/2} \rightarrow {}^4I_{13/2}$ transition are crowded by so many lines (Figure 2).

Question arises about the assignment of each of the observed lines to a well defined transition between a given sublevel of the ground $^4I_{15/2}$ to one of the excited $^4I_{13/2}$ manifold, respectively. The correct attribution allows building the energy sublevel scheme for every manifold. The task can be accomplished if the spectra 1) are measured at high resolution (at low temperatures), 2) their temperature dependence is followed at rather close steps (a beautiful example is portrayed in Figure 7), 3) the RE^{3+} concentration in the sample is rather low to minimize the spectral line broadening and overlapping and to avoid the presence of additional lines, due to the RE^{3+}-RE^{3+} interaction, see Figure 3 and Sec. III, and 4) the experimental energy levels are fitted by using a suitable crystal field model (Sec. II C).

1. An Example of Line Attribution: Er^{3+} in Crystalline Environments

The line attribution on the basis of the experimental data is henceforward illustrated for the Er^{3+} $^4I_{15/2} \rightarrow {}^4I_{13/2}$ transition in BaYF by following [10]. The choice of Er^{3+} is based 1) on the large number of manifolds (41), thus it represents a general example rather than the end member ions as Ce^{3+} and Yb^{3+} (Sec. II B 3) and 2) in spite of the large number of lines (Figure 2), the procedure is simpler than for the Dy^{3+}, Ho^{3+}, and Tm^{3+} (Sec. II B 2).

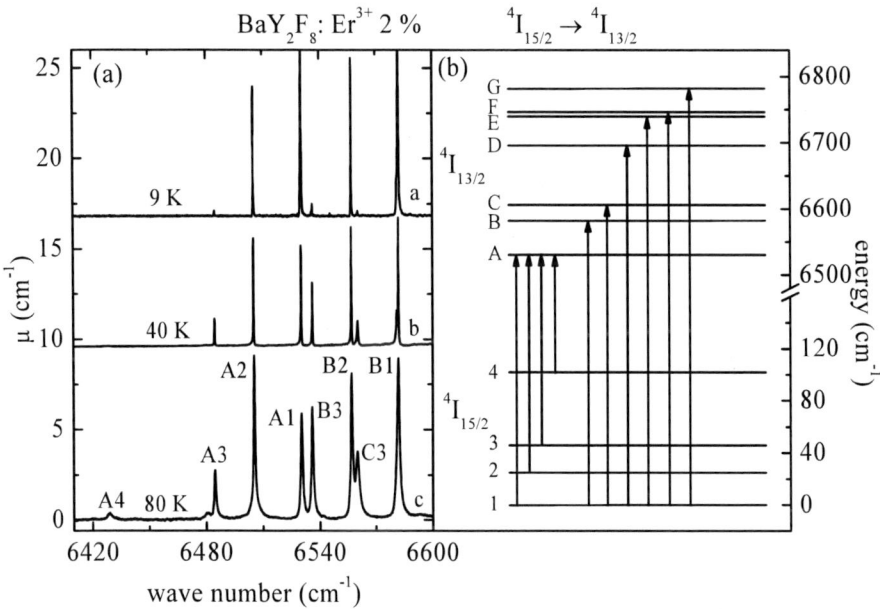

Figure 4. Panel (a): optical absorption spectra measured at different temperatures in a limited range of $^4I_{15/2} \rightarrow {}^4I_{13/2}$ transition (6410-6600 cm^{-1}) for a BaY_2F_8 sample doped with 2 at% Er^{3+}. Curve a: 9 K, curve b: 40 K, curve c: 80 K. Panel (b): crystal field splitting of the ground $^4I_{15/2}$ (only the first four sublevels are displayed for clarity) and excited $^4I_{13/2}$ Er^{3+} manifolds. Some possible absorption transitions are indicated by arrows.

If the spectra are measured at rather low temperatures (for example T=9 K), the narrow absorption lines, see curve a in Figure 4(a), are due to transitions from the lowest sublevels (the only ones which are populated) of the ground manifold $^4I_{15/2}$ to all the sublevels of the excited manifold $^4I_{13/2}$, see Figure 4(b). By increasing the temperature, even higher sublevels of the ground manifold $^4I_{15/2}$

become gradually populated. As a consequence, 1) the line amplitude changes, compare in Figure 4(a) curve b (T=40 K) with curve a (T=9 K) and 2) the number of transitions (and, as a consequence, of absorption lines) to different sublevels of the excited manifold $^4I_{13/2}$ increases. This effect is clearly displayed by Figure 4(a): the 80 K spectrum (curve c) is more complex than that taken at 9 K (curve a). Thus the careful analysis of the spectra, as a function of the temperature, is the key for supplying a correct attribution of the absorption lines and for evaluating the ground and excited manifold splitting, as illustrated by Figures 4 and 5. The line at 6530.23 cm^{-1} (labeled as A1) is strong at 9 K and decreases monotonically by increasing the temperature, compare curves a, b, and c in Figure 4(a). The lines at 6505.15 cm^{-1} (A2), at 6484.36 cm^{-1} (A3), which are both already detectable at 9 K, and at 6428.60 cm^{-1} (A4), which is absent at 9 K, increase by increasing the temperature, compare curves a, b, and c in Figure 4(a). As the temperature increases the lines broaden: a detailed plot of the A2 and A3 line amplitude vs. T shows that after an initial increase the amplitude decreases at T>50 K, see Figure 7 in Ref. [10]. It is more meaningful to plot the areas subtended to the lines (proportional to the line strengths): the result is that the areas under the A2, A3, and A4 lines monotonically increase with the temperature, see Figure 7 in Ref. [10]. To extend the above described analysis to a different host (YAB) and to a different Er concentration, Figure 5 displays the temperature dependence of A1 (at 6527 cm^{-1}) and A2 (at 6480 cm^{-1}) line amplitude [Panel (a)] and area [Panel (b)]. In this case the low Er concentration (0.3 mol%) allows to measure correctly even the area under the A1 line: such a task cannot be accomplished by using the BaYF: Er^{3+} (2 at%) data, because the A1 line amplitude is too large at low temperatures, thus it is either out of scale [see for example Figure 4(a) curve a] or affected by the MCT detector non-linearity [10]. Due to the low dopant level in YAB: Er^{3+} (0.3 mol%), the A3 line amplitude (and as a consequence the area) could not be plotted vs. T, because it was negligible: on the contrary it was easily detected for higher Er concentrations [9,11,18,19].

The above results suggest that the A1 line is originated by the transition starting from the lowest sublevel of the ground manifold $^4I_{15/2}$ to the lowest sublevel of the excited manifold $^4I_{13/2}$, in fact the lowest sublevel of $^4I_{15/2}$ [labeled as 1 in Figure 4(b)] must be populated at 9 K. By increasing the temperature, its population decreases in favor of the sublevels 2, 3, 4, ... of the $^4I_{15/2}$ manifold [Figure 4(b)]. Transitions may occur from these sublevels to the lowest sublevel of the excited manifold $^4I_{13/2}$, originating the A2, A3, and A4 lines, whose area increases by increasing the temperature, see Figure 7 in Ref. [10] and compare curves a, b, and c in Figure 4(a).

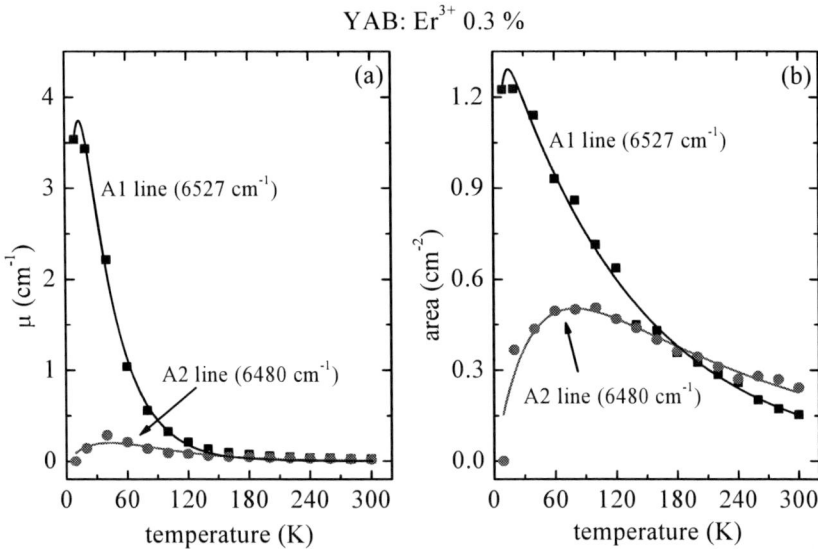

Figure 5. Temperature dependence of the amplitude and area of two lines (A1 and A2) belonging to $^4I_{15/2} \rightarrow {}^4I_{13/2}$ transition for a YAB: Er^{3+} 0.3 mol% sample. Panel (a): amplitude vs. temperature for A1 (6527 cm^{-1}, black squares) and A2 (6480 cm^{-1}, red circles) lines. Panel (b): area subtended to A1 (black squares) and A2 (red circles) lines as a function of the temperature. The lines are guide for the eyes.

The separation between the A1 line position and that of A2, A3, and A4 lines gives the energy separation of the sublevels 2, 3, and 4 from the lowest lying sublevel 1 in the ground manifold $^4I_{15/2}$, respectively, see Figure 4(b). The positions of the eight sublevels of the ground manifold $^4I_{15/2}$ can be evaluated in the same way by monitoring the spectra at different temperatures starting from 20 K, by 20 K steps. At 180 K all the $^4I_{15/2}$ sublevels are already populated and the corresponding lines could be easily identified in the spectrum, while at higher temperatures the line broadening and overlapping make the line identification and attribution quite difficult, see for example curves a and b in Figure 1.

The procedure, here explained for the Ai lines (i=1, 2,..., 8), was repeated to identify all the lines in the spectrum, i.e. the lines due to transitions from the different sublevels of the ground manifold (i=1, 2, ..., 8) to different sublevels of the excited manifold (e.g. A, B, C, D, E, F, and G for the $^4I_{13/2}$ manifold), see Figure 4(b). Each line was labeled as Xi, where X=A, B, C, ... indicates a given sublevel of an excited manifold and i=1, 2, ..., 8 the sublevel of the ground manifold, from which the absorption transition starts, see Figure 4(b). As examples, Figures 2 a-f and 4(a) display some of the Xi lines identified for the

$^4I_{15/2} \to {}^4I_{13/2}$ transition. The energy values of the $^4I_{15/2}$ and $^4I_{13/2}$ manifold sublevels are displayed in Table 2. The procedure was extended to all the investigated transitions: i.e. $^4I_{15/2} \to {}^4I_{11/2}$, $^4I_{9/2}$, $^4F_{9/2}$, $^4S_{3/2}$, $^2H_{11/2}$, $^4F_{7/2}$, and $^4F_{5/2}$ and the sublevels energies were determined [10,20].

Table 2. Sublevel positions (cm^{-1}) for Er^{3+} ground ($^4I_{15/2}$) and first excited ($^4I_{13/2}$) manifolds in different matrices, as derived from optical absorption measurements performed at different temperatures

$^{2S+1}L_J$	Sublevel	BaY$_2$F$_8$[a]	YAB[b]	YAG[c,d]	ErF$_3$[e]	Er$_2$O$_3$[e]	SnO$_2$[f]
$^4I_{15/2}$	1	0	0	0	0	0	0
	2	25.01±0.08[g]	47.06	22	90	39.54	46
	3	45.6±0.7[h]	109	61	110.4	75.28	68.0
	4	98.5±0.9[h]	128	79	131.7	88.96	85.0
	5	280.5±3.9[h]	157	-	154.8	-	-
	6	329.8±4.4[h]	244	-	176.7	-	-
	7	362.0±4.1[h]	287	-	194.3	-	-
	8	400.3±1.9[h]	316	-	249.8	-	-
$^4I_{13/2}$	A	6530.23	6527	6547.5	6583.8	6511.9	6533.3
	B	6582.44	6561	6597.1	6606.8	6533.4	6561.2
	C	6606.23	6611	6603.8	6665.7	6543.7	6568.6
	D	6695.60	6639	6784.1	6686.3	6595.6	6571.3
	E	6739.10	6724	6803.6	6719	6688.5	6586.2
	F	6746.19	6735	-	6735	6842.2	6609.6
	G	6781.45	6740	6883.3	6794.2	6869.2	6724.1

[a] Ref. [10]
[b] Ref. [11]
[c] Ref. [1]
[d] Ref. [21]
[e] Ref. [17]
[f] present work.
[g] measured at 9 K.
[h] measured at 180 K.
Each sublevel is labeled, in the second column, with either a number or a letter, see text.

Similar approach was followed to identify the Er^{3+} energy level scheme in YAB single crystals [9,11,18,19] and in Er$_2$O$_3$ and ErF$_3$ polycrystalline powders [17]. Thanks to the wide transparency window and good optical quality of Er-doped BaYF and YAB single crystals, a large number of Er^{3+} transitions could be investigated; on the contrary the spectra of Er$_2$O$_3$ and ErF$_3$ polycrystalline powders could be measured only thanks to their inclusion in KBr pellets: the light scattered by the powders precluded any spectral analysis in the high wave number region (i.e. higher than 10400 cm^{-1}) [17]. The Er^{3+} sublevel positions within the $^4I_{15/2}$ and $^4I_{13/2}$ manifolds are collected in Table 2 and compared for different matrices.

The number of sublevels for each excited manifold was found to correspond to the expected value of (2J+1)/2 for symmetries lower than the cubic one. In the case of polycrystalline samples it was not always possible to identify all the sublevels of the ground manifold for the reasons listed above.

2. Line Attribution for More Complex Spectra (Dy^{3+}, Ho^{3+}, and Tm^{3+})

The procedure to build the energy level scheme, above exemplified for Er^{3+} in BaYF, should be in principle easily extended to other trivalent RE^{3+}, as Dy^{3+}, Ho^{3+}, and Tm^{3+}, which have 9, 10, and 12 electrons in the 4f shell, respectively. However the line attribution may be not simple if two sublevels are rather close one to each other either in the ground or in the excited manifold.

Dy^{3+} is a Kramers ion, as Er^{3+}, but the separation between the first two sublevels of the $^6H_{15/2}$ ground manifold is 7.58 and 3.3 cm^{-1} in BaYF and YAB single crystals, respectively, see Table 3 and Refs. [11,22,23], to be compared with the 25.01 and 47.06 cm^{-1} for Er^{3+} in the same matrices, see Table 3. This means that the two sublevels are almost equally populated at low temperature (e.g. 9 K) and the lines (X1 and X2) originated by transitions starting from them have comparable intensity. Thus the task to determine the correct separation between sublevels 1 and 2 within the $^6H_{15/2}$ ground manifold is not so easy. For the narrowest lines, as for example those lying at the lowest energies within the $^6H_{15/2} \rightarrow {^6H_{13/2}}$ and $^6H_{11/2}$ transitions, respectively, the problem could be solved by exploiting the high resolution and the detailed analysis of their temperature dependence [11].

A complementary approach was also followed. Linear dichroism measurements allowed identifying two close sublevels (E and F) within the $^6H_{11/2}$ excited manifold, notwithstanding their separation was only 1.6 cm^{-1} and the related E1 and F1 lines in unpolarized spectra overlapped giving rise to a rather broad band peaking at ~6046.9 cm^{-1}, see curve a in Figure 6(a) [11]. The broad band splits into two peaks at 6047.97 and 6046.73 cm^{-1} if the measurements are performed with light electric field E$_l$ parallel and perpendicular to z axis, see curves b and c, respectively in Figure 6(a).

Therefore linear dichroism measurements work as an additional tool for a correct line attribution in spectral ranges were the unpolarized spectra are rather complex, because they reveal the symmetry of states between which the transition occurs (Sec. II B 5).

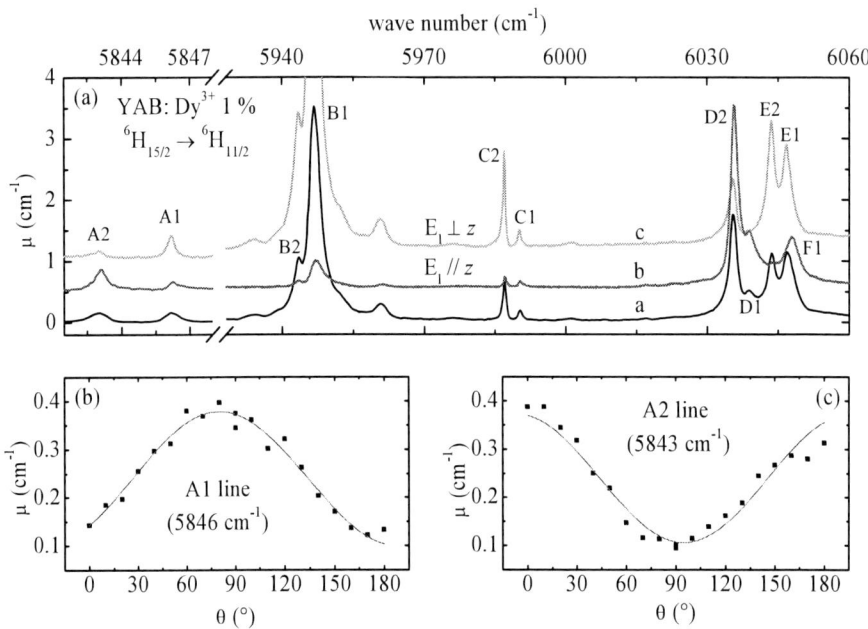

Figure 6. Absorption spectra of a YAB: Dy^{3+} 1 mol% sample in the region of $^6H_{15/2} \to {}^6H_{13/2}$ transition measured at 9 K with polarized and unpolarized light. Curve a: unpolarized spectrum; curve b: light electric field E_l parallel to z axis; curve c: light electric field E_l orthogonal to z axis. Panels (b) and (c) show the polarization dependence of the absorption coefficient for A1 (5846 cm^{-1}) and A2 (5843 cm^{-1}) lines, respectively. The solid lines are only guides for the eyes.

The line attribution for non-Kramers ions, as Ho^{3+} and Tm^{3+}, with an even number of 4f electrons (10 and 12, respectively), is still more complex due to the large number of sublevels (up to 2J+1) into which a manifold may be split by low symmetry crystal field. For example, the lowest two manifolds $^5I_8$ and $^5I_7$ of Ho^{3+}, embedded in monoclinic BaYF, may split into 17 and 15 sublevels, respectively. Such a splitting originates a large number of possible transitions between the sublevels of the two manifolds giving rise to very crowded line spectra [24]. The extremely small separation (0.6 cm^{-1}) between the two first sublevels of the $^5I_8$ ground manifold, see Table 3, adds further difficulties, as above explained for Dy^{3+} in BaYF and YAB. For Ho^{3+} in YAB the first two manifold degeneracy is not completely lifted by the trigonal crystal field (space group R32). Thus the number of lines originated by the transitions between the sublevels of the $^5I_8$ and $^5I_7$ manifolds is lower than for Ho^{3+} in BaYF. However further complications arise due to 1) hyperfine splitting of some levels (Sec. III D and Figure 22) and 2) small separations between sublevels 2 and 3 (2.07 cm^{-1}) of the ground $^5I_8$ manifold and between sublevels A and B (2.2 cm^{-1}) of the first excited $^5I_7$ manifold [25].

Table 3. Crystal field splitting of the first two sublevels within the ground manifold for different RE^{3+} ions embedded in BaY$_2$F$_8$ and YAB matrices

Matrix	BaY$_2$F$_8$					YAB		
Dopant	Ce^{3+}	Dy^{3+}	Ho^{3+}	Er^{3+}	Tm^{3+}	Dy^{3+}	Ho^{3+}	Er^{3+}
X1-X2	80	7.58	0.6	25.01	1.6	3.3	12.26	47.06
σ_{ST}		0.52	0.06	0.08	0.1	0.2	0.35	0.82
N		35	28	29	9	26	86	25

The separations X1-X2 (cm^{-1}) are estimated averaging the difference between the wave numbers of several pairs (N, last raw) of X1 and X2 lines. The last but one row reports the standard deviation σ_{ST} (cm^{-1}).

For Tm^{3+} in monoclinic BaYF, the number of lines originated by the transitions between the lowest two manifolds $^3H_6$ and $^3F_4$ is not as high as for Ho^{3+} in the same matrix, but the two first sublevels of the ground $^3H_6$ manifold are separated only by 1.6 cm^{-1} [24], see Table 3. In spite of the above listed difficulties, it was possible to provide the correct attribution of so many lines in Ho^{3+} doped BaYF and YAB and in Tm^{3+} doped BaYF [24,25], thanks to the high resolution, the extended spectral range investigated, the detailed line temperature dependence, and the accurate theoretical analysis of the experimental data, see Sec. II C. An example of the close-spaced temperature dependence for the Tm^{3+}-doped BaYF spectra in the region of the $^3H_6 \rightarrow {}^3F_4$ transition is portrayed in Figure 7. For clarity the spectra are displayed over a limited wave number range (5330-5730 cm^{-1}) in Figure 7(a). A further magnification of the line peaking at 5683.5 cm^{-1} (A1 line) is provided by Figure 7(b). As expected for a X1 line, its amplitude decreases by increasing the temperature. The shoulder appearing at ~5682 cm^{-1}, clearly detectable in the 9 K spectrum [Figure 7(b)], is the A2 line, which reveals the small separation between the 1 and 2 sublevels of the ground manifold. By increasing the temperature A1 and A2 lines merge, broaden and their intensity decreases in favor of other *hot lines*, i.e. originated by transitions starting from excited sublevels of the ground $^3H_6$ manifold, as clearly shown on the left by Figure 7(a).

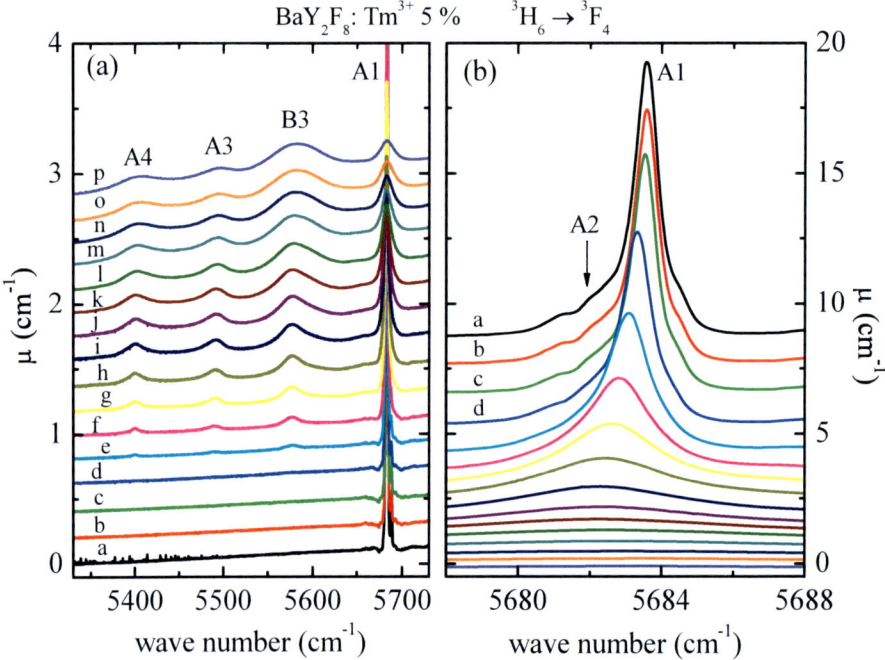

Figure 7. Optical absorption spectra measured at different temperatures in the region of Tm^{3+} $^3H_6 \to ^3F_4$ transition on a BaY_2F_8: Tm^{3+} (5 at%) sample. The zero-phonon transitions are labeled with Xi (X=A, B,... and i=1, 2,...). Curve a: 9 K; curves from b to p: temperature varying from 20 to 300 K by 20 K steps. Panel (a): 5330-5730 cm^{-1} range to show the growth of *hot lines*; panel (b): magnification of the 5678-5688 cm^{-1} range to show the A1 line temperature dependence.

An additional help to the correct line attribution comes from measuring the spectra on samples in which the Tm^{3+} concentration is quite small, see Figure 8, where the 9 K spectra related to Tm^{3+} 5 and 0.5 at% concentration, respectively, are compared in a wave number range similar to that of Figure 7(b). In Figure 8 the A2 line, although very weak, is clearly separated from A1 in curve a (low concentration) at variance with curve b (high concentration), where additional lines, indicated by dot-dashed arrows (Sec. III B), and line broadening make more difficult the line identification. On the other hand, by measuring only low concentration samples, some weak lines are missed. Thus a sound level attribution can be pursued also by analyzing samples with different dopant concentrations.

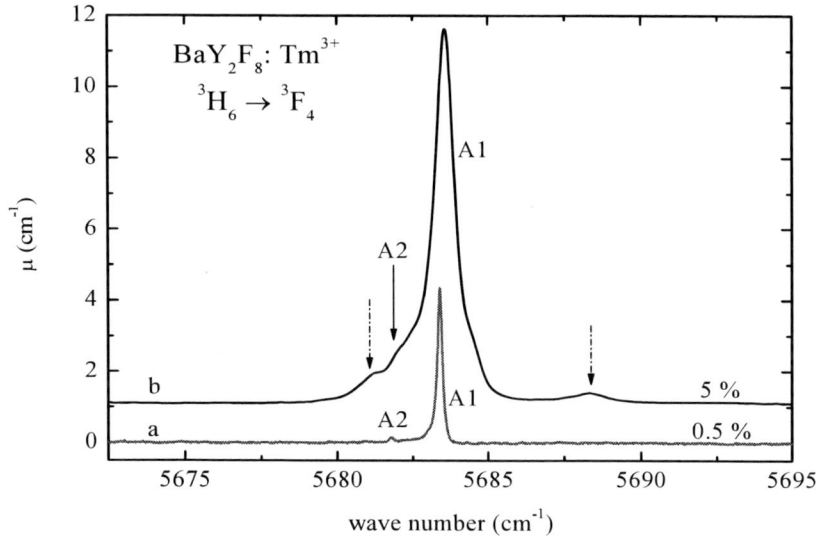

Figure 8. Magnification of the optical absorption spectra measured at 9 K in the region of Tm^{3+} $^3H_6 \rightarrow {}^3F_4$ transition for BaY_2F_8 single crystals doped with different Tm^{3+} concentrations. Curve a: 0.5 at%; curve b: 5 at%. The dot-dashed arrows indicate the additional lines.

3. Line Attribution for End Members of the RE Series: The Example of Ce^{3+}

For completeness, the much simpler spectrum of an end member of RE^{3+} series, as Ce^{3+}, is depicted in Figure 9: the whole spectrum induced by the 4f-4f transitions of Ce^{3+} is confined in the IR (2100-3500 cm^{-1} range), due to the unique $^2F_{5/2} \rightarrow {}^2F_{7/2}$ transition. By analyzing the spectra temperature dependence [Figure 9(a)], it is easy to build the energy level scheme [Figure 9(b)]. However, due to the limited total number of levels (7), the fitting of their experimental values cannot be as good as for other RE^{3+}, characterized by a larger number of levels and, as a consequence, by more complex spectra. However, under some reasonable assumptions important information can be extracted from these simpler spectra as well (Sec. II C).

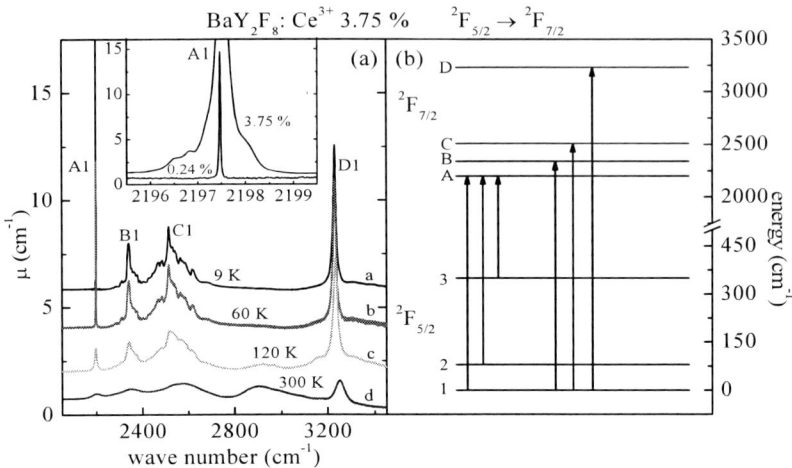

Figure 9. Panel (a): optical absorption spectra measured at different temperatures in the region of Ce^{3+} $^2F_{5/2} \rightarrow {}^2F_{7/2}$ transition for a BaY_2F_8 sample doped with 3.75 at% Ce^{3+}; curve a: 9 K, curve b: 60 K, curve c: 120 K, curve d: 300 K. The inset shows a magnification of the optical absorption spectra measured at 9 K in the region of A1 line for BaY_2F_8 single crystals doped with different Ce^{3+} concentrations: 0.24 at% and 3.75 at%, respectively (res. 0.02 cm^{-1}). Panel (b): crystal field splitting of the ground $^2F_{5/2}$ and excited $^2F_{7/2}$ Ce^{3+} manifolds. Some possible absorption transitions are indicated by arrows.

4. Sublevel Separation within the Ground Manifold

Thanks to the wide wave number range investigated in the case of 9 K high resolution absorption spectra of BaYF and YAB single crystals doped with different RE^{3+} (Dy^{3+}, Er^{3+}, Ho^{3+}, Tm^{3+}), it was possible to evaluate with high accuracy the positions of the first sublevels of the ground manifold with respect to the position of sublevel 1, which was assumed to lye at zero energy: the separation Δ_{n-1} between the positions of Xn and X1 lines could be averaged for X=A, B, C, ... in a given manifold and over all the manifolds investigated. For example in the case of BaYF: Er^{3+} the position of the sublevel 2 was estimated from the difference Δ_{2-1} between the wave numbers of 29 pairs of X1 and X2 lines in BaYF. The mean value was 25.01±0.08 cm^{-1} at 9 K [10], which should be compared with the value 26±0.5cm^{-1} at 77 K obtained from fluorescence measurements performed at a resolution of 0.5 cm^{-1} [26]. The accuracy decreases by increasing the temperature at which the measurements are performed, although high resolution spectra covering a wide wave number region are available, due to line broadening and overlapping, see Table 2, third column, were the high sublevel position within the ground $^4I_{15/2}$ state was determined at relatively high temperature.

Even for systems affected by the above listed difficulties, the adopted procedure allowed estimating with high accuracy the separation even between very close sublevels within the ground manifold. Examples are collected in Table 3. The most striking one is $\Delta_{2-1}=0.6\pm0.06$ cm^{-1} for Ho^{3+} in BaYF: at least to our knowledge for the first time so tiny separation is detected [27]; in fact, previously reported fluorescence measurements supplied a value of 20 cm^{-1} for Δ_{2-1} [1,28], a separation which has not been observed in the present high resolution absorption spectra.

Table 4. Electric-and magnetic-dipole selection rules for D_3 symmetry

D_3	ED		MD	
	Γ_4	$\Gamma_{5,6}$	Γ_4	$\Gamma_{5,6}$
Γ_4	π, σ	σ	σ, π	π
$\Gamma_{5,6}$	σ	π	π	σ

As a rule, fluorescence measurements may provide easily the full set of the ground manifold sublevel positions, however by monitoring high resolution absorption spectra as a function of temperature, in most cases it was possible to reach the same goal, with much higher accuracy: the Ho^{3+} $^5I_8$ manifold sublevel identification in YAB represents a nice example [25].

5. Linear Dichroism Measurements and Wavefunction Symmetry

As reported in Sec. II B 2, polarized light spectra allowed separating absorption lines which are close and indistinguishable in the unpolarized spectrum, see Figure 6(a). The result is based on the symmetry of the states involved in a given transition and on the selection rules for a specific transition (electric or magnetic dipole allowed). For example, the A1 and A2 lines within the $^6H_{15/2} \rightarrow {}^6H_{11/2}$ transition of Dy^{3+} in YAB, whose amplitude is practically the same in the unpolarized spectrum [Figure 6(a), curve a], have different statistical weights if the spectra are measured with polarized light, i.e. with light electric field E_l either along the z axis (curve b) or perpendicular to it (curve c). The opposite trend followed by A1 and A2 lines is displayed in detail by Figures 6(b) and 6(c), where the related amplitudes are plotted vs. the angle θ between E_l and z axis. Similar plots have been obtained for most of the lines belonging to the

$^6H_{15/2} \rightarrow ^6H_{13/2}$, $^6H_{11/2}$, $^6H_{9/2}+^6F_{11/2}$, $^6H_{7/2}+^6F_{9/2}$, and $^6F_{7/2}$ transitions. Lines were labeled with π and σ, if the amplitudes exhibited their maxima for light polarized along z axis and perpendicularly to it, respectively. The results were analyzed by using the D_3 symmetry selection rules [6] to determine the polarization character of the electric- and magnetic- dipole transitions, by keeping in mind that the crystal field states of Dy^{3+} are either Γ_4 or $\Gamma_{5,6}$ [29], see Table 4.

Table 5. Polarization character for each transition related to the absorption lines observed in the linear dichroism measurements performed on a YAB: Dy^{3+} 1 mol% sample

Transition	Line	ν (cm^{-1})	χ_P	Transition	Line	ν (cm^{-1})	χ_P
$^6H_{15/2} \rightarrow ^6H_{13/2}$	A1	3570.98	π	$^6H_{15/2} \rightarrow ^6H_{9/2}+^6F_{11/2}$	G1	7844.44	σ
	B1	3579.76	(π σ)		H1	7875.53	σ
	C1	3598.18	σ		L1	7963.87	π
	D1	3644.07	π		M1	7984.06	σ
	E1	3712.48	σ		N1	8023.50	σ
	F1	3769.82	σ	$^6H_{15/2} \rightarrow ^6H_{7/2}+^6F_{9/2}$	A1	9047.14	π
	G1	3810.41	(π σ)		B1	9069.86	σ
$^6H_{15/2} \rightarrow ^6H_{11/2}$	A1	5846.18	σ		C1	9111.41	(π σ)
	B1	5946.72	σ		D1	9189.50	σ
	C1	5990.2	σ		E1	9192.75	π
	D1	6038.77	π		F1	9244.78	π
	E1	6046.73	σ		G1	9320.71	σ
	F1	6047.96	π		H1	9335.47	σ
$^6H_{15/2} \rightarrow ^6H_{9/2}+^6F_{11/2}$	A1	7692.21	σ		L1	9348.93	π
	B1	7713.90	(π σ)	$^6H_{15/2} \rightarrow ^6F_{7/2}$	A1	11042.9	σ
	C1	7736.10	σ		B1	11093.3	π
	D1	7756.92	σ		C1	11094.5	σ
	E1	7810.72	π		D1	11143.7	σ
	F1	7819.00	σ				

As a first approximation, the magnetic dipole (MD) contribution was assumed to be negligible. The approximation is acceptable mainly for terms endowed by low J. According to the selection rules for electric dipole (ED) applied to the observed X1 lines (Sec. II B 1) belonging to the $^6H_{15/2} \rightarrow ^6F_{7/2}$ transition, the lowest state of the ground manifold was identified as a $\Gamma_{5,6}$. On the basis of this result and of the selection rules it was possible to predict the polarization character χ_P of each X1 line for the $^6H_{15/2}\rightarrow{}^6H_{13/2}$, $^6H_{11/2}$, $^6H_{9/2}+{}^6F_{11/2}$, $^6H_{7/2}+{}^6F_{9/2}$, and $^6F_{7/2}$ transitions and compare it with the experimental values collected in Table 5 [30]. The agreement was quite good, except for a few discrepancies indicated within brackets. Such differences may be ascribed to having neglected the MD contribution for high J values and to some local distortion of the hexagonal symmetry around Dy^{3+}. Once more the high resolution absorption spectra measured at low T provide, in the framework of linear dichroism measurements aimed to get insight on the Dy^{3+} state symmetry in YAB, an ample set of clear-cut results, which cannot be accomplished from polarized spectra taken at 300 K by using dispersive spectrometers [31]. The opposite trend vs. θ displayed in Figure 6(b) and 6(c) by A1 and A2 line at ~5846 and 5843 cm^{-1}, respectively, is now easily understood by considering the different symmetry of states 1 ($\Gamma_{5,6}$) and 2 (Γ_4) from which the transitions, originating the two lines, start.

C. CRYSTAL FIELD ANALYSIS, ENERGY LEVEL SCHEME, AND SUPERPOSITION MODEL

In the well-known Hartree model, each electron in an atom is treated as if it were moving independently in a net potential of spherical symmetry, describing as an average the Coulomb interaction with the nucleus and with the other electrons. Historically, this very simple description has progressively been made more accurate by taking into account several perturbative potentials which represent other (weaker) interactions involving the electrons belonging to partially occupied shells (in the case of trivalent rare earths, only the 4f electrons play a role). The two most important free-ion terms are the residual electron-electron Coulomb repulsion

$$H_{ee} = \sum_{i>j=1}^{n} \frac{e^2}{r_{ij}} \tag{1}$$

and the spin-orbit interaction

$$H_{so} = \zeta \sum_i \mathbf{s}_i \cdot \mathbf{l}_i \tag{2}$$

where s_i and l_i are the spin and orbital angular momentum of the i-th electron.

It can be shown that the operators representing the total orbital momentum $\mathbf{L} = \sum_i \mathbf{l}_i$ and the total spin $\mathbf{S} = \sum_i \mathbf{s}_i$ commute with H_{ee}, hence its eigenstates are arranged into multiplets (*terms*) labeled by the quantum numbers L and S. Inside a single, well-isolated term the spin-orbit coupling can be expressed as

$$H'_{so} = \Lambda \mathbf{L} \cdot \mathbf{S} \tag{3}$$

and it is diagonal within the basis composed of the eigenvectors of the total angular momentum $J = L + S$, hence giving rise to several manifolds conventionally labeled as $^{2S+1}L_J$.

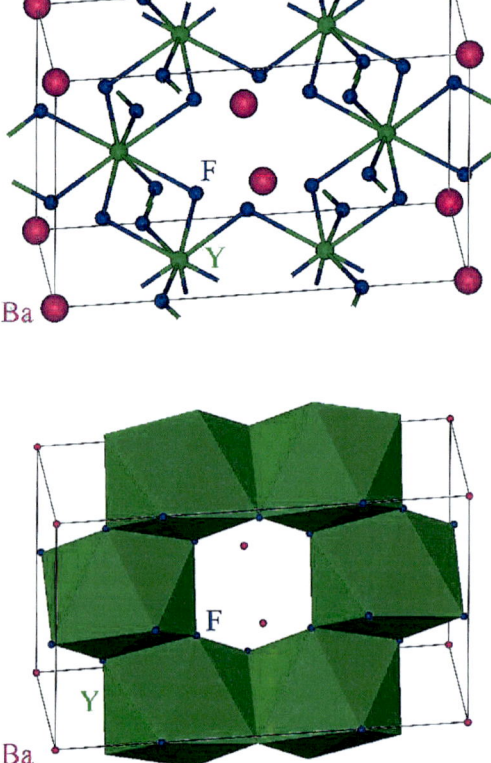

Figure 10. Unit cell of BaY_2F_8 where bonds between the elements (top) and the F^- polyhedra (bottom) are put in evidence. Each cell contains two formula units.

Each manifold is (2J+1)-fold degenerate in a free ion due to the spherical symmetry of the Hamiltonian. When placing the ion in a crystal this symmetry is significantly lowered and each energy level splits under the influence of the electric field produced by the environment, see, for example, in Figures 10-13 the surroundings tested by RE^{3+} in the crystal hosts analyzed in the present work. Of course, the free-ion degeneracy should not be expected to be completely removed; this is possible only for RE^{3+} ions with an even number of f electrons placed in a crystal site with sufficiently low point symmetry. As a general consideration, the energy levels of an odd-electron RE^{3+} ion in a crystal are at least doubly degenerate unless a magnetic field (or another time-invariance-breaking interaction) is present (Kramers' theorem).

The basic hypothesis behind the crystal field (CF) theory is that the electrons of the incomplete shell remain localized on the ion when it is placed in the crystal, so that the crystal field potential H_{CF} can be considered as a perturbation on the free-ion energy levels. Since we are interested in rare-earth ions, this hypothesis is generally valid, given the small average radius of the 4f shell around the nucleus. H_{CF} can then be written in terms of spherical harmonics [32]

$$C_q^{(k)} = \sqrt{\frac{4\pi}{2k+1}} \sum_i Y_{kq}(\theta_i, \varphi_i) \tag{4}$$

as

$$H_{CF} = \sum_{k=2,4,6} \sum_{q=-k}^{k} B_k^q C_q^{(k)} \tag{5}$$

where the B_k^q coefficients are the so-called crystal field parameters and the matrix elements of tensor operators can be obtained as an application of the Wigner-Eckart theorem:

$$\langle \alpha, L, S, J, M | C_q^{(k)} | \alpha', L', S', J', M' \rangle = \delta_{SS'} (-1)^{k+S+L'+2J-M+l} (2l+1) \begin{pmatrix} l & k & l \\ 0 & 0 & 0 \end{pmatrix}$$
$$\times \sqrt{(2J+1)(2J'+1)} \begin{pmatrix} J & k & J' \\ -M & q & M' \end{pmatrix} \begin{Bmatrix} L & J & S' \\ J' & L' & k \end{Bmatrix} \langle \alpha, L, S \| U^{(k)} \| \alpha', L', S' \rangle$$
$$\tag{6}$$

with $l = 3$ for f electrons and using the reduced matrix elements $\langle \alpha, L, S \| U^{(k)} \| \alpha', L', S' \rangle$ tabulated by Nielson and Koster [33]. The number of nonzero crystal field parameters depends on the local symmetry of the rare-earth site(s) in the crystal.

The Coulomb, spin-orbit, and crystal field interactions give the right order of magnitude for the energy level splittings of $4f^n$ configurations. However, these terms cannot accurately reproduce the experimental data alone, and further corrective potential must be added. These additional terms will be shortly reviewed in the following, referring to the original papers for a more detailed treatment.

An accurate calculation of the optical properties of a rare-earth ion should, in principle, take into account the non-diagonal matrix elements between states with different configurations. Approaching this problem is far from being a simple task: apart from the huge matrices which would be involved in such an algorithm, the effective operators defined so far retain their validity only within a fixed configuration. Luckily, it is possible to introduce a set of effective operators which reproduce the effects of a multi-configurational calculation while acting within the ground $4f^n$ configuration. The two-body configuration interaction is written as [32]

$$H_2 = \alpha G(R_3) + \beta G(G_2) + \gamma G(R_7) \tag{7}$$

where α, β, and γ are the so-called Trees parameters and $\hat{G}(X)$ represents the eigenvalues of Casimir's operators for the X group [recalling that $\hat{G}(R_3) = L(L+1)$]. For $4f^n$ configurations with $3 \leq n \leq 11$ a three-body interaction term was also introduced [34]:

$$H_3 = \sum_{i=2,3,4,6,7,8} T^i t_i \tag{8}$$

In addition to the magnetic spin-orbit interaction parametrized by ζ, relativistic effects including spin-spin and spin-other-orbit interaction are included in the Hamiltonian by means of Marvin parameters M^j [35]:

$$H_3 = \sum_{j=0,2,4} M^j m_j \tag{9}$$

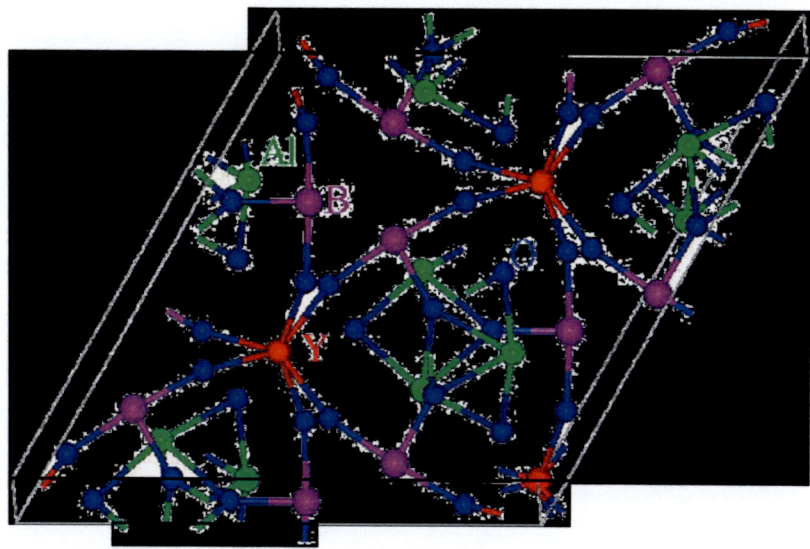

Figure 11. The unit cell of YAB structure where bonds between the elements are put in evidence. Each cell contains three formula units.

Finally, the fitting of experimental energy levels can be improved by introducing two-body effective operators to account for configuration interaction through electrostatically correlated magnetic interactions:

$$H_5 = \sum_{k=2,4,6} P^k p_k \qquad (10)$$

where the three additional parameters P^k have to be considered [36].

The free-ion Hamiltonian for a given rare earth is generally considered to be very similar in all phases; therefore, we generally use the parameters given by Carnall et al. [37] for doped LaF_3 as starting values. (Carnall's analysis is particularly valuable as the determined values are in agreement with expected systematic trends [38], so that care should be taken if the final values are too dissimilar from these.) On the other hand, the crystal field parameters are not expected to change drastically when different rare earths are embedded into the same host. A good example is rare-earth-doped YAB. The trigonal-prismatic 6-fold oxygen coordination (Figure 11) around the rare earth site results in a D_3 local symmetry, which allows only nonzero B_k^q real parameters with $q = 0,3,6$ in the crystal field Hamiltonian. The experimental energy levels for trivalent Dy, Ho, and Er in YAB were then fitted by varying the crystal field, Coulomb, and spin-orbit parameters [11]. This minimal choice was made to avoid over-parameterization, since the energy levels within the experimentally reachable range belong to only a few different ^{2S+1}L terms; the other weaker corrective potentials were kept fixed. The resulting free-ion parameters are very similar to the starting values and follow reasonable trends with increasing atomic number; on the other hand, the crystal field parameters remain more or less constant along the series (Table 6). While the crystal field symmetry is low enough to split the energy manifolds of the Kramers ions Dy^{3+} and Er^{3+} into the maximum allowed number of levels, i.e. $(2J+1)/2$ doublets, the spectra of Ho^{3+} is composed by singlets (of A_1 and A_2 symmetry) and doublets (of E symmetry) [25,39]. Nevertheless, in both cases the crystal field calculations can determine the symmetry properties of the wavefunctions corresponding to each energy level; this can give useful information since, for example, electric or magnetic dipole transitions may be forbidden between levels of certain symmetries. This behaviour can be tested, for example, by polarized light experiments, see Sec. II B 5.

Table 6. Coulomb, spin-orbit and crystal field parameters determined for various trivalent rare-earth ions in YAB

	Dy^{3+}	Ho^{3+}	Er^{3+}
F^2	91025	93675	96329
F^4	63864	66430	68001
F^6	49462	51767	53342
ζ	1904.7	2140.8	2369.6
B_2^0	505	491	530
B_4^0	-1495	-1150	-1297
B_6^0	283	327	214
B_4^3	-814	-797	-632
B_6^3	-75	-62	-97
B_6^6	-244	-162	-175

All values are in cm^{-1}.

The most important feature of the crystal field theory is that its validity can be extended well beyond the boundaries of the electrostatic point-charge model [40]; the symmetry-allowed parameters in the crystal field Hamiltonian can be determined by fitting experimental data. Once they are known, useful information on the active site and its surroundings can be extracted by ab-initio or semiphenomenological models. A well-known example is Newman's Superposition Model. The two main assumptions of this model are that:

i) The CF potential is the sum of individual contributions from each ligand (in general limited to nearest neighbours).
ii) These contributions are cylindrically symmetric, the symmetry axis being the straight line which links the considered ligand to the central ion.

In this framework, the CF parameters can be expressed as

$$B_k^q = N_k^q \langle r^k \rangle \sum_l \overline{A}_k(R_l) K_k^q(\theta_l, \varphi_l) \qquad (11)$$

where $K_k^q(\theta, \varphi)$ are the coordination factors defined in Ref. [41], l labels the ligands and $\overline{A}_k(R_l)$ is an intrinsic parameter, whose distance dependence is usually supposed to follow the power law

$$\overline{A}_k(R) = \overline{A}_k(R_0) \left(\frac{R_0}{R}\right)^{t_k} \qquad (12)$$

with t_k and $\overline{A}_k(R_0)$ as adjustable parameters and R_0 as a conveniently defined "standard ligand distance" [38]. One should note that the point-charge model can be derived from these more general expressions by fixing suitable values of t_k and $\overline{A}_k(R_0)$ [41]. On the other hand, Eq. 5 retains its validity as long as the f-electron wavefunctions maintain their essentially localized character, e.g. it correctly accounts for strongly covalent bonding and it can also be used to describe the quantum state of RE^{3+} ions in 3d-4f intermetallic compounds.

To be useful for interpretation, the transferability postulate is usually invoked. Quoting Newman's own words [41], it states that "*single-ligand contributions are dependent only on the nature of the ligand and its distance from the paramagnetic ion, and do not depend on other properties of the host crystal*". Again, this is straightwardly valid in a point-charge framework and means that t_k and

$\overline{A}_k(R_0)$ should be more or less the same for the same ligand in a reasonable distance range around R_0.

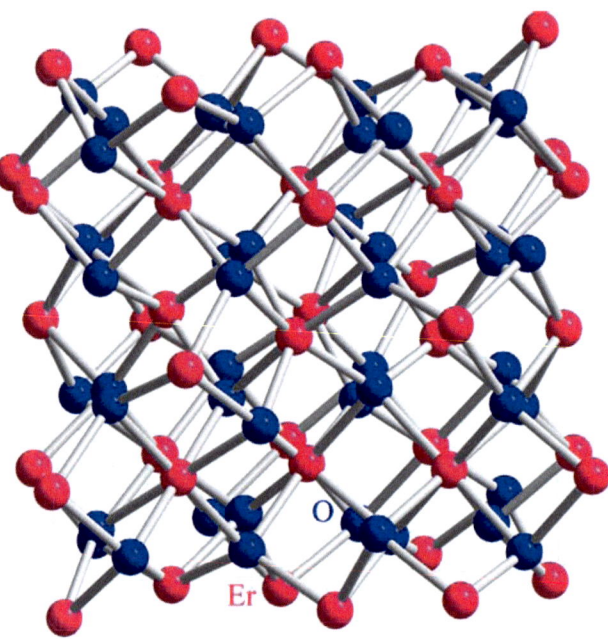

Figure 12. The unit cell of Er_2O_3 structure where bonds between the elements are put in evidence. Each cell contains four formula units.

This type of analysis cannot resolve the exact electronic structure of the magnetic center, but it extracts definite information on the correlation of the RE^{3+} energy levels with the geometry around the active site. For example, free-ion and crystal field parameters were determined for several different rare earths in BaYF (Table 7). In this case, the complicated structure (Figure 10) leads to a lower local symmetry (C_2) and as much as fourteen non-zero crystal field parameters appear (B_2^1 can be forced to zero by fixing the reference frame). As in the case of YAB, free-ion trends are respected and the crystal field potential does not vary excessively along the series. We then fitted the so-obtained crystal field values for R = Er, Dy, and Nd with the superposition model (Table 8): good results were achieved with the same values of t_k and $\overline{A}_k(R_0)$, but assuming that the nearest-neighbour F^- polyhedron is slightly compressed for heavier dopants, as a result of their lower atomic radius [42].

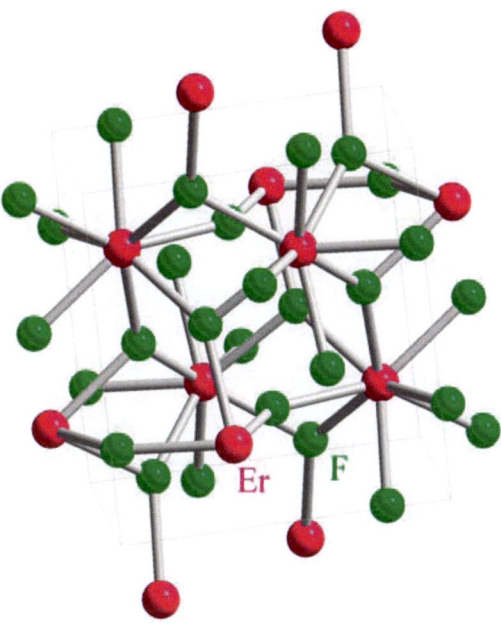

Figure 13. The unit cell of ErF$_3$ structure where bonds between the elements are put in evidence. Each cell contains four formula units.

If reliable parameters for a given ligand-dopant pair are already known, the superposition model can be useful to estimate approximate values for the crystal field parameters in different hosts. The resulting potential can even be used to calculate the energy spectra directly, then eventually fitting t_k and $\overline{A}_k(R_0)$ to the experimental levels. This "reverse" approach can be suitable in cases where too little experimental information is known (or can be known) to reasonably fit all the necessary free parameters in the single-ion Hamiltonian. As mentioned before, the Ce^{3+} spectra is only composed of 7 doublets, 3 belonging to the lowest $^2F_{5/2}$ and 4 to the excited $^2F_{7/2}$ manifold (Sec. II B 3). The only relevant free-ion parameter is the spin-orbit coupling (as only one term is present), but the number of crystal field parameters to be considered can be quite large. We have applied this method to the case of Ce-doped BaYF, obtaining very good results with the parameters given in Table 8; the values of $\overline{A}_4(R_0)$ and $\overline{A}_6(R_0)$ are in line with those of the other rare earths, as predicted by the transferability principle. $\overline{A}_2(R_0)$, on the other hand, is quite smaller: this has been attributed to an expansion of the F⁻ ligand cage, and to the non-negligible long-range electrostatic contribution from Ba^{2+} next-to-nearest neighbours [23].

Table 7. Coulomb, spin-orbit and crystal field parameters determined for various trivalent rare-earth ions in BaYF

	Nd^{3+}	Dy^{3+}	Ho^{3+}	Er^{3+}	Tm^{3+}
F^2	72625	90000	94042	96354	100061
F^4	53086	65060	66541	68601	69368
F^6	35425	48267	51464	53204	55451
ζ	880	1911	2142	2362	2631
B_2^0	-620	-500	-527	-450	-443
B_4^0	-1123	-1580	-1301	-1430	-1113
B_6^0	780	650	463	470	472
B_2^2	-120	-10	70	60	59
B_4^1	240	450	319	350	272
B_4^2	60	60	100	110	86
B_4^3	700	-130	4	5	4
B_4^4	470	440	373	410	319
B_6^1	-190	-230	-247	-250	-251
B_6^2	10	50	-49	-50	-50
B_6^3	-240	-20	-79	-80	-80
B_6^4	890	430	385	390	392
B_6^5	-610	-390	-365	-370	-372
B_6^6	-90	100	247	250	251

All values are in cm^{-1}.

Table 8. Superposition model parameters obtained for various systems

Compound	BaY_2F_8	BaY_2F_8	BaY_2F_8	BaY_2F_8	ErF_3	Er_2O_3
Ligand-dopant	Ce^{3+} -	Nd^{3+} -	Dy^{3+} -	Er^{3+} - F⁻	Er^{3+} -	Er^{3+} -
$\overline{A}_2 (R_0)/cm^{-1}$	30	630	720	720	510	670
$\overline{A}_4 (R_0)/cm^{-1}$	22	23	48	48	20	51
$\overline{A}_6 (R_0)/cm^{-1}$	3.15	2.5	3.8	3.8	2.6	4.6
t_2	5	5	5	5	5	3
t_4	6	6	6	6	6	10
t_6	10	10	10	10	10	11
$R_0/\text{Å}$	2.275	2.275	2.275	2.275	2.275	2.275

A similar analysis was performed for absorption spectra of ErF$_3$ and Er$_2$O$_3$. Er$_2$O$_3$ has a complex crystal structure (Figure 12) belonging to the space group Ia3. Each erbium is situated at the centre of a cube of oxygens from which two oxygens of the nearest neighbours are removed. Two arrangements of the remaining oxygens occur: in one case the oxygens are removed from a cube body diagonal producing a C$_{3i}$ point symmetry, while in the other arrangement the two oxygens are removed from a cube face diagonal, thus the Er^{3+} site has a C$_2$ symmetry. ErF$_3$, as all lanthanide fluorides, exhibits an orthorhombic symmetry with space group Pnma. Each erbium ion is surrounded by eight fluorine ions at comparable distances and a ninth one at larger distance. Six of the nine F ions form a trigonal prism around the Er^{3+} ions, while the remaining three are in front of the lateral faces, so that the polyhedron is a tricapped prism (Figure 13). Measurements were taken on CsI-diluted pellets, therefore in a limited energy range: only two excited manifolds (^{4}I$_{13/2}$ and ^{4}I$_{11/2}$) were visible [17]. Despite the complex crystal field potential and the paucity of experimental data, it was possible to extract all the superposition model parameters (Table 8), which could eventually be used to calculate the energy levels for erbium in different crystalline environments and unravel the complex spectra of erbium in nanostructured glassceramics, by obtaining information on the ligand geometry around the optically active ions (Sec. III A 3).

Chapter III

RE^{3+} AS A PROBE OF THE ENVIRONMENT

A. LINE WIDTH AS A PROBE OF THE RE^{3+} ENVIRONMENT

As a rule the absorption lines due to 4f-4f transitions of RE^{3+} in crystals are very narrow due to the shield provided to 4f electrons by the 5s and 5p shells. This feature is quite evident in high resolution spectra measured at low temperature. Examples are portrayed by curves a-f in Figure 2 for the lowest $^4I_{15/2} \rightarrow ^4I_{13/2}$ transition of Er^{3+} in different crystalline matrices. As a general remark, the lines which lie at lower wave numbers are sharper than those peaking at high wave numbers. A similar behaviour is observed in the spectral region of 1) the unique 4f-4f transition ($^2F_{5/2} \rightarrow ^2F_{7/2}$) of Ce^{3+} in BaYF, see Figure 9, curve a, 2) the lowest transition ($^6H_{15/2} \rightarrow ^6H_{13/2}$) of Dy^{3+} in YAB, see Figure 1 in Ref. [11], and 3) the lowest transition ($^5I_8 \rightarrow ^5I_7$) of Ho^{3+} in BaYF, see Figure 1 in Ref. [24]. Such a trend is emphasized for lines resulting from transitions having as final states the sublevels of higher energy manifolds. For example, in BaYF, for a fixed Er^{3+} concentration (2 at%), the width of A2 lines peaking at 6505.16 ($^4I_{15/2} \rightarrow ^4I_{13/2}$), 10202.27 ($^4I_{15/2} \rightarrow ^4I_{11/2}$), and 19143.63 cm^{-1} ($^4I_{15/2} \rightarrow ^4H_{11/2}$) is 0.1, 0.53, and 1.6 cm^{-1}, respectively [10]. This increased line width can be in part explained by the opening of a larger number of spontaneous emission decay channels for the high lying levels: this involves the level lifetime shortening, and, as a consequence, the line broadening [10]. An additional relevant source of line broadening in low temperature spectra is supplied by the inhomogeneous one, due to disorder in the host crystalline matrix, see Sec. III A 1.

By increasing the temperature the line width increases, see for example Figures 4(a), 7(b), and 9(a) for Er^{3+} ($^4I_{15/2} \rightarrow ^4I_{13/2}$), Tm^{3+} ($^3H_6 \rightarrow ^3F_4$), and Ce^{3+}

($^2F_{5/2}\rightarrow{}^2F_{7/2}$) in BaYF, respectively, due to the interaction with the lattice modes (homogeneous broadening), see Sec. IV A.

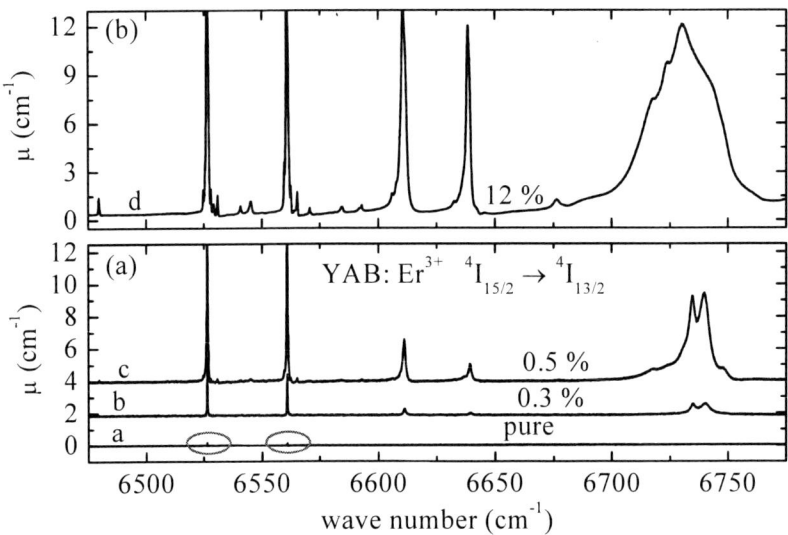

Figure 14. Absorption spectra of Er^{3+} in YAB in the region of $^4I_{15/2}\rightarrow{}^4I_{13/2}$ transition. Curve a: YAB: Ho^{3+} 1 mol% with traces of Er^{3+}; curve b: YAB: Er^{3+} 0.3 mol%; curve c: YAB: Er^{3+} 0.5 mol%; curve d: YAB: Er^{3+} 12 mol%. All the spectra are measured at 9 K. The absorption line of Er^{3+} detected in the Ho^{3+} doped sample (curve a) are marked with circles.

1. Long Range Order Perturbed by Increasing RE^{3+} Concentration

The position of the main lines in 9 K spectra of a given system is almost independent of the RE^{3+} concentration, but not their width, as displayed by Figures 2 (curves a and b) and 14 for Er^{3+} ($^4I_{15/2}\rightarrow{}^4I_{13/2}$ transition) in BaYF and YAB, respectively. A magnification of a restricted wave number range in both cases [see Figures 3 and 15(a), respectively], for Tm^{3+} and Ce^{3+} in BaYF [see Figure 8 and inset in Figure 9(a), respectively] supplies a still more convincing proof. For low RE^{3+} concentration the lines are very sharp: for example the width of the A1 line at 2197 cm^{-1} due to 0.24 at% Ce^{3+} in BaYF is less than 0.03 cm^{-1}. In samples where Er^{3+} is present only as a trace the lines are still narrower: the width of the A1 line at 6530.23 cm^{-1} in BaYF is below 0.02 cm^{-1} (to be compared with 0.13 cm^{-1} for 0.5 at% concentration) [10] and that of the B1 line at 6561 cm^{-1} in

YAB is ~0.04 cm^{-1} [to be compared with 0.2 cm^{-1} for 0.3 mol% concentration, see curves a and b in Figure 15(a)]. The results are along with that reported for the 15302.4 cm^{-1} line in a low-strain, isotopically pure (99.9% ^{7}Li) LiYF$_4$ single crystal containing naturally occurring levels of Er^{3+} impurities (~1 p.p.m.): the line widths are in the 10^{-3}-4×10^{-3} cm^{-1} range [43]. In this case the inhomogeneous broadening, responsible for the very small line width, is due to ^{167}Er^{3+} hyperfine splitting and to Er^{3+} isotopes present in different natural abundances.

The line width vs. the nominal RE concentration cannot be easily analyzed if some additional lines overlap, as shown in Figure 15(a). The task can be correctly accomplished by choosing a line which remains isolated even at high RE^{3+} doping level. A good choice is offered by the A3 Er^{3+} line at 6484.36 cm^{-1} in BaYF, whose width increases by a factor 13 if the Er content grows from 0.5 to 20 at%, see Figure 15(b) and Ref. [10].

RE^{3+} line broadening at low temperature is tightly connected to the disorder probed by the RE^{3+}: a clear example is represented by the Er^{3+} spectra in glassy SiO$_2$: the very narrow lines detected in crystals as BaYF, YAB, and YAG (curves a-d in Figure 2) merge in a broad band when Er^{3+} is embedded in silica, see curve i in Figure 2, due to structural disorder of the glassy matrix.

The RE^{3+} concentration induced line broadening can be explained by considering that the random distribution of RE^{3+} over the Y^{3+} sites reduces the long range order within the crystal host. The RE^{3+} 4f electrons, in spite of the shield provided by the outermost 5s and 5p electrons, experience slightly different crystal fields, whose intensity and symmetry depend on other RE^{3+} distance and distribution. By increasing the RE^{3+} concentration, the average distance among the RE^{3+} decreases and RE^{3+}-RE^{3+} loose interactions give rise to inhomogeneous line-broadening. This is confirmed by the line shape change at high dopant levels. In RE^{3+} diluted solid solutions the narrowest lines are Lorentzian shaped, as shown clearly in Figure 15(a) by the red curve which fits according to a Loretzian curve b, related to the Er^{3+} 6561 cm^{-1} line in YAB. By increasing the concentration, the shape, even of isolated lines, as that peaking at 6484.36 cm^{-1} in BaYF, turns to a Voigt profile or to a Gaussian, see Figure 3 in Ref. [10], due to a random distribution of slightly different surroundings monitored by Er^{3+} [44]. Similar effects were monitored for very narrow OH stretching lines in KMgF$_3$, whose line width increases and shape changes by increasing the Pb^{2+} content [45]. A line broadening involves excited state lifetime shortening which was detected indeed in BaYF by increasing Er^{3+} content, see Table 4 in Ref. [10].

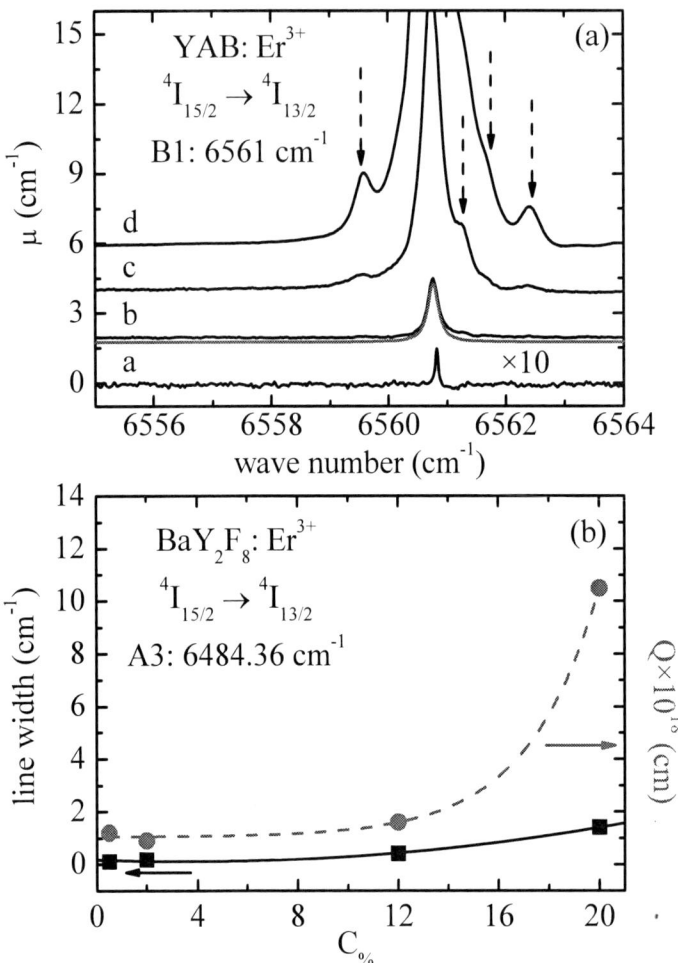

Figure 15. Panel (a): optical absorption spectra at 9 K related to B1 line at ~6561 cm^{-1} within the $^4I_{15/2} \to {}^4I_{13/2}$ transition of Er^{3+} in YAB. Curve a: YAB: Ho^{3+} 1 mol% with traces of Er^{3+}; curve b: YAB: Er^{3+} 0.3 mol%; curve c: YAB: Er^{3+} 0.5 mol%; curve d: YAB: Er^{3+} 12 mol%. The ordinate of curve a is multiplied by a factor 10. The dashed arrows put in evidence the additional lines. All spectra were measured at 0.04 cm^{-1} resolution.
Panel (b): line width and oscillator strength vs. Er^{3+} concentration for A2 line at ~6484 cm^{-1} within Er^{3+} $^4I_{15/2} \to {}^4I_{13/2}$ transition in BaY_2F_8: Er^{3+} samples. Black squares: line width; red circles: oscillator strength (Q). Black solid and red dashed lines are only guides for the eyes.

The analysis of a given line (e.g. A3 line at 6484.36 cm^{-1} in BaYF) oscillator strength Q (i.e. the ratio between the area subtended to the line and the concentration of Er^{3+} ions in the sample) vs. the nominal Er^{3+} concentration reveals an interesting feature displayed in Figure 15(b): Q remains practically constant, as expected, for low dopant concentrations, but grows according to a nearly exponential trend for high doping levels. Usually a superlinear dependence of line intensity vs. the dopant concentration, allows attributing the line to dopant clusters, but this is not the case because the 6484.36 cm^{-1} peak is a *hot line* (A3) which appears even in diluted Er^{3+} solid solution, therefore it should be attributed to isolated Er. The impressive Q rise for Er 20 at% can be explained as a change of the Er^{3+} transition probability, due to a local crystal field symmetry lowering, caused by the presence of other rather close Er ion(s) [10].

2. Short Range Order Induced by Fluorine in Er^{3+} Doped Silica Glasses

As mentioned in Sec. III A 1, the structural disorder typical of glassy networks, as that of silica, broadens to a great extent the Er^{3+} lines even in the 9 K spectra, otherwise very sharp in crystalline hosts. In Figure 16, in spite of rather low Er^{3+} content (0.5 mol%) in a sol-gel silica sample, the 9 K spectrum (curve a) shows only two broad bands: these can be considered an envelope of the two groups of lines detected in some Er^{3+} doped crystals, compare, for example, curve i with curves a-d in Figure 2. To minimize the OH content in Er doped SiO$_2$ (for optical applications based on RE^{3+} radiative emissions) the sol-gel synthesis is performed by starting from fluorinated precursors. OH, responsible for undesired luminescence quenching [46,47], can be dislodged by fluorine in two ways. Fluorine may either react with Si-OH groups giving Si-O bonds and HF molecules, which leave the glass during the thermal treatment, or substitute OH producing a network terminating Si-F groups [14]. The fluorine remaining in the glass causes interesting ordering effects which are monitored by Er^{3+}, which acts as a probe. A 2 mol% fluorinated precursor concentration (Sec. II A) induces impressive changes in the region of the Er^{3+} $^4I_{15/2} \rightarrow {}^4I_{13/2}$ transition. The fluorinated sample spectrum (curve b in Figure 16) exhibits sharp lines, superimposed to a residual weak broad structure, reminiscent of that displayed by curve a. The Er^{3+} line sharpening is practically complete if the fluorinated precursor concentration is raised to 5 mol%, see curve c in Figure 16. By increasing the temperature the sharp lines exhibit homogeneous broadening as for Er^{3+} in an ordered surrounding (Sec. IV A). The result suggests that in fluorinated silica Er^{3+} probes environments, which are either ordered, at least locally, or partially decoupled from the structurally disordered glassy network [14]. Fluorine may play indeed a twofold role: 1) to release stresses in silica [48] and 2) to interact with Er^{3+}, which easily substitutes Si in Si-F terminating groups, which are already in part decoupled from the glassy network [49]. The number of Er^{3+} lines displayed by curves b and c in Figure 16 is much larger than that expected even in the case of the lowest crystal field symmetry (Sec. II B). The large amount of lines may arise from different local arrangements: one or more non-bridging F^- may substitute for one or more O^{2-} around Er^{3+} [14]. The presence of crystalline ErF_3 separate phase, which in principle might be responsible for the narrow lines, must be ruled out, because none of the lines detected coincides with those displayed by microcrystalline ErF_3 powder spectrum, see curve f in Figure 2.

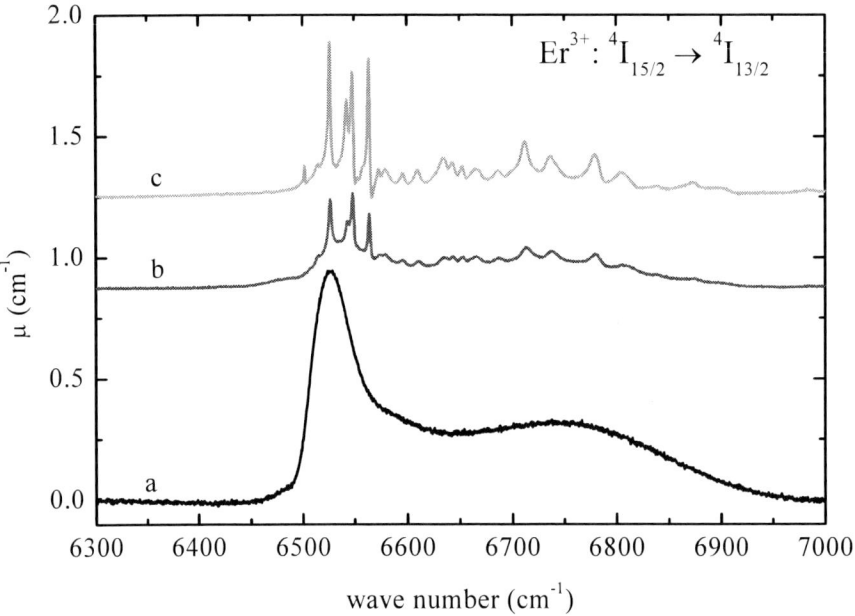

Figure 16. Optical absorption spectra measured at 9 K in the region of Er^{3+} $^4I_{15/2} \rightarrow ^4I_{13/2}$ transition for three sol-gel silica samples. Curve a: SiO_2: Er^{3+} 0.5 mol%; curves b and c: SiO_2: Er^{3+} 0.2 mol% prepared from fluorinated precursors (F 2 and 5 mol%, respectively).

3. Local Order Monitored by Er^{3+} in Nanostructured SiO_2/SnO_2 Glassceramics

Quantum-Dots (QDs) systems are of large interest for telecommunications: an example is represented by transparent glasses, compatible with silica based technology, in which crystalline nano-phases, endowed with non-linear optical properties, are embedded. Proper thermal treatments, applied to silica doped with tin in concentrations higher than 1 mol%, cause the segregation of tin-dioxide (SnO_2) particles of nanometric dimensions [12], giving rise to glass-ceramic bulk samples with a quite narrow cluster size distribution [13]. SnO_2 is a good candidate for wide-band-gap QDs system since it has an energy-gap of 3.6 eV and a rather high third order non-linear optical coefficient [50]. SnO_2 nano-crystallites, characterized by low phonon energies ($E_{p,max} \approx 670$ cm^{-1}), are suitable hosts for luminescent rare earth ions, as for example Er^{3+}. In principle, the Er^{3+} doped SiO_2/SnO_2 glass-ceramic matches the requirements to design optical devices in which the Er^{3+} laser emission and non linear optical properties are simultaneously exploited.

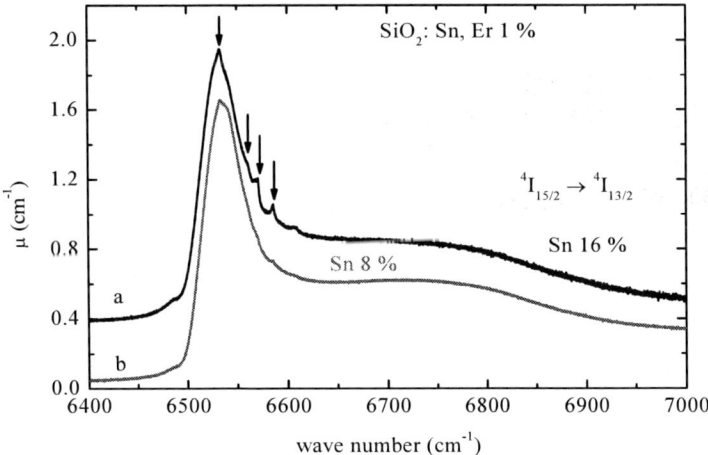

Figure 17. Optical absorption spectra measured at 9 K in the region of Er^{3+} $^4I_{15/2} \rightarrow {}^4I_{13/2}$ transition for two glass-ceramic samples doped with 1 mol% Er^{3+}. Curve a: SiO_2: SnO_2 16 mol%; curve b: SiO_2: SnO_2 8 mol%. The solid arrows put in evidence the lines which are present even in Er^{3+} doped SnO_2 samples.

Although the crystalline nature of SnO_2 nanoclusters within the glass-ceramic can be identified by means of transmission electron microscopy (TEM) and electron diffraction analysis [13,15], no TEM image can distiguish Er^{3+} ions randomly dispersed in amorphous silica from those embedded in a crystalline nanocluster. This task is successfully accomplished by the high resolution absorption spectroscopy applied at low temperature, because Er^{3+} line width works as a sensitive probe of the RE^{3+} surrounding. As already mentioned in Sec. III A 2, the low temperature spectra of Er^{3+} doped silica are broad and poorly structured, see curve a in Figure 16, because Er^{3+} probes the disordered glass network. The RT spectrum of an Er^{3+} doped SiO_2/SnO_2 glass-ceramic in the region of the $^4I_{15/2} \rightarrow {}^4I_{13/2}$ transition resembles quite closely that of Er^{3+} doped silica, compare curve h to curve i in Figure 1, but if high resolution measurements are performed at 9 K, weak, narrow lines overlap the broad structure, see curve h in Figure 2 and curves a and b in Figure 17. This means that some of the Er^{3+} ions monitor an ordered environment as in a crystal, while most of them are located in disordered surroundings. Ordered environments are available indeed in SnO_2 nanocrystals. The narrow line amplitude increases by increasing the Sn concentration in the glass, compare curves a and b in Figure 17. This suggests that Er^{3+} has more ordered sites to visit, because there are larger SnO_2 nanocrystals and/or a larger number of nanoclusters. However the direct proof of Er^{3+} inclusion in SnO_2 nanocrystals comes from the comparison with the spectra of microcrystalline Er^{3+} doped SnO_2, as portrayed in Figure 2. The weak, narrow lines at ~6533, 6559, 6571, 6586 cm^{-1}, detected in Er^{3+} doped glass-ceramic (Figure 2, curve h and Figure 17, curves a and b) coincide with those monitored in Er^{3+} doped SnO_2, see Figure 6(b) in Ref. [15]. Broad structures are present even in the Er^{3+} doped SnO_2 spectrum (Figure 2, curve g and Figure 17): this means that Er^{3+} probes disordered surroundings, possibly in partially amorphous SnO_2 and/or at grain boundaries in the polycrystalline sample. Such broad structures can be envisaged even in the glass ceramic spectrum (Figure 2, curve h) in addition to those typical of the Er^{3+} doped silica (Figure 2, curve i and Figure 16, curve a), if curve h is fitted to a superposition of gaussian-shaped bands. Thus in the glass ceramic sample, Er^{3+} may be embedded in SnO_2 nanocrystals (narrow lines) and in amorphous phases (broad bands), as glassy silica and disordered SnO_2 arrangements, which may occur, for example, at the interface between the tetragonal crystalline SnO_2 and the surrounding tetrahedrally coordinated glass.

If the Er^{3+} doped glass ceramic is synthesized starting from fluorinated precursors (Sec. II A) with the aim to reduce the OH content (Sec. III A 2) many more narrow lines overlap the broad spectrum, see Figure 18, curve a (9 K): the lines indicated by arrows at ~6533, 6559, 6571, 6586 cm^{-1}, attributed to Er^{3+} in ordered SnO_2 sites are still present in addition to those induced by the ordering and network decoupling role played by F (Sec. III A 2). Other narrow lines arise possibly due to Er^{3+} probing environments in which Sn and F are present. By temperature increasing the narrow lines are characterized by a homogeneous broadening, see Figure 18, curves a-d, a feature typical of RE^{3+} embedded in an ordered environment, at variance with the behaviour monitored in glasses where the spectra structure remains broad even by lowering the temperature, compare curves i in Figures 1 and 2, respectively.

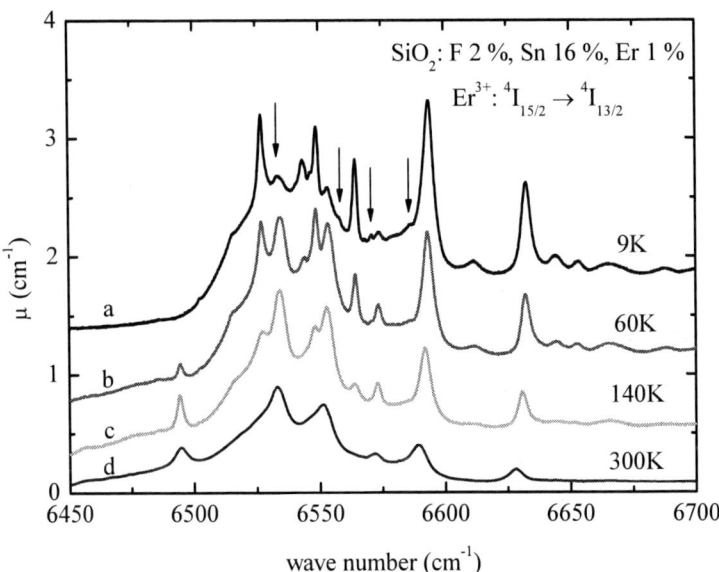

Figure 18. Optical absorption spectra measured at different temperatures in the region of Er^{3+} $^4I_{15/2}\rightarrow\,^4I_{13/2}$ transition for a SiO_2: SnO_2 16 mol%, Er^{3+} 1 mol% glass-ceramic sample obtained from fluorinated precursors (F 2 mol%). Curve a: 9 K, curve b: 60 K, curve c: 140 K; curve d: 300 K.

B. RE^{3+} CLUSTERS

As described in Sec. III A 1, by increasing the RE^{3+} concentration the lines, attributed to the RE^{3+} dispersed in the crystal matrix, broaden and their oscillator strengths change, see Figure 15(b). Besides these effects, additional weak lines appear in the 9 K high resolution spectra, if the dopant concentration is rather high, see Figures 3, 8, 9 (insert), and 15(a). The presence of additional lines is not peculiar of a specific matrix or RE, but is a general feature, as suggested by the above mentioned figures, related to Er^{3+}, Tm^{3+}, and Ce^{3+} doped BaYF and Er^{3+}

doped YAB, respectively. These lines are weak if compared to the main lines, related to isolated RE^{3+}. An example is provided by comparing Figure 19 with Figure 2, curve a. In the former, which gives a magnified presentation, four lines are clearly distinguishable, while in the latter the same lines are merged in the background (compare the ordinate scales). Furthermore, for a given system, the additional lines accompany different Xi lines, originated by transitions starting from the sublevels of ground manifold to those of many excited manifolds [10]. None of the additional lines finds an explanation from crystal field calculations for an isolated RE^{3+} dispersed in the crystal host. Their amplitude exhibits a superlinear dependence on the RE^{3+} concentration. A detailed study has been carried out in the case Er^{3+} in BaYF [10]: a given Xi line was accompanied by a set of additional lines (labeled as $\alpha_1, \alpha_2, ..., \alpha_n$) which covered about 20 cm^{-1}, mainly on the low wave number side of the reference Xi line. The association of an α line set to a given Xi line was based on a similar dependence of the α_n and Xi lines on the temperature and on the light polarization, see Figure 8 in Ref. [10], Sec. II B 1, and Sec. II B 5, respectively. An example of the same dependence on the polarization angle θ for a few α lines belonging to the same set which accompanies the A1 line at 6530.2 cm^{-1} in BaYF: Er^{3+} (Figure 19) is displayed in Figure 20. Thus the α lines were attributed to Er^{3+} sitting in sites characterized by a crystal field slightly modified with respect to that probed by an isolated Er^{3+}, due to a more or less loose interaction with one or more close Er^{3+}.

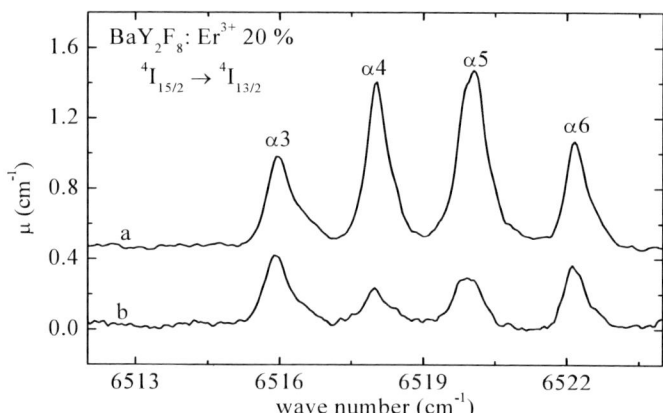

Figure 19. Dichroic optical absorption spectra measured at 9 K on a BaY_2F_8: Er^{3+} 20 at% sample in the region of Er^{3+} $^4I_{15/2} \to {}^4I_{13/2}$ transition where the additional lines occur. Curve a: light electric field parallel to the [002] axis; curve b: light electric field orthogonal to the [002] axis.

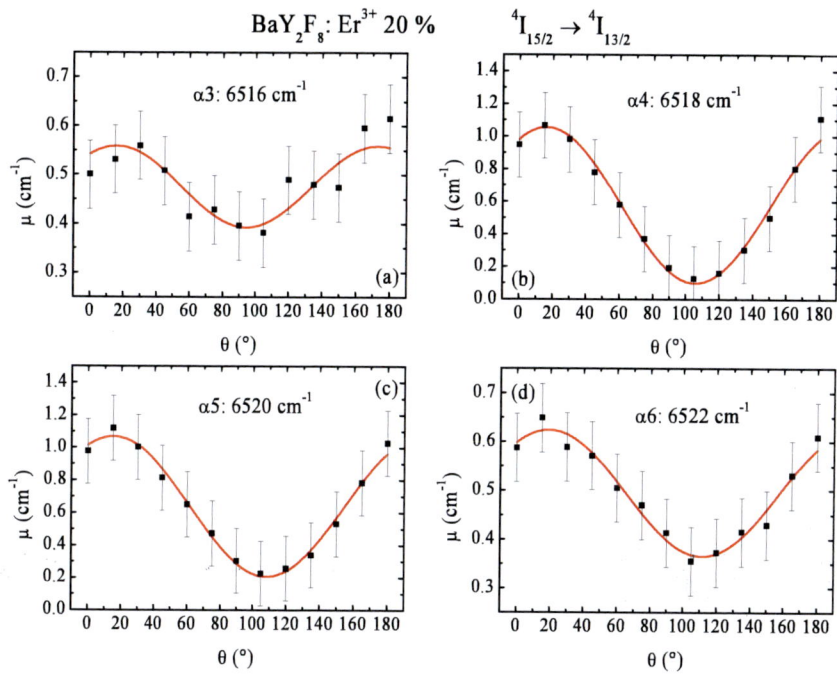

Figure 20. Polarization dependence of the absorption coefficient for the additional lines exhibited by a BaY$_2$F$_8$: Er^{3+} 20 at% sample in the region of Er^{3+} $^4I_{15/2} \rightarrow {}^4I_{13/2}$ transition. Panel (a): α3 line at 6516 cm^{-1}; panel (b): α4 line at 6518 cm^{-1}; panel (c): α5 line at 6520 cm^{-1}; panel (d): α6 line at 6522 cm^{-1}. The solid lines are only guides for the eyes.

The conclusion is supported by the presence of α lines in BaYF samples with high Er^{3+} doping level, e.g. 12 and 20 at%, see Figure 4(b) in Ref. [10]: in a 20 at% Er^{3+} doped sample every five Y^{3+} sites one is occupied by an Er^{3+}. Two (or three) Er^{3+} may occupy near neighboring (nn) or next near neighboring (nnn) YF$_8$ polyhedra, see Figure 21, giving rise to a fine tuning of the Er^{3+}-Er^{3+} loose interaction [20], and as a consequence to a series of Er^{3+}-lines slightly shifted from the stronger lines due to isolated Er^{3+}. In spite of the α line weakness, for a few of them it was possible to supply at least a partial level scheme, in the sense that the differences Δ_{2-1} (between the first two sublevels in the ground manifold, see Sec. II B 4), Δ_{B-A}, and Δ_{C-A} (between the first two and between the third and the first sublevels, respectively in a few excited manifolds, see Sec. II B 1) could be easily identified, see Table 7 in Ref. [10].

The above mentioned Δ are quite close to those found for isolated Er^{3+} ions, supporting the hypothesis of a weak Er^{3+}-Er^{3+} interaction. The presence of a BaEr$_2$F$_8$ separated phase was excluded by x-ray diffraction measurements [10].

Calculations based on the Newman's superposition model provide a rough estimate of the shifts, which are consistent with the observed ones. By correlating the uncertainty on the fitted crystal field parameters to a possible slightly off-center position of the rare-earth ions substituting yttrium, we found that a reasonably good agreement is obtained within a distortion of about 0.05 Å, not easily detectable by standard diffraction techniques in diluted compounds.

The presence of the additional lines for relatively high RE^{3+} doping levels stresses once more the sensitivity of the RE^{3+} in probing the surrounding, notwithstanding the shield provided to 4f electrons first of all by the outermost 5s and 5p shells and also by the eight fluorines of the YF_8 polyhedra and by borate groups interposition in BaYF [51] and YAB [52], respectively, see Figure 10 and 11.

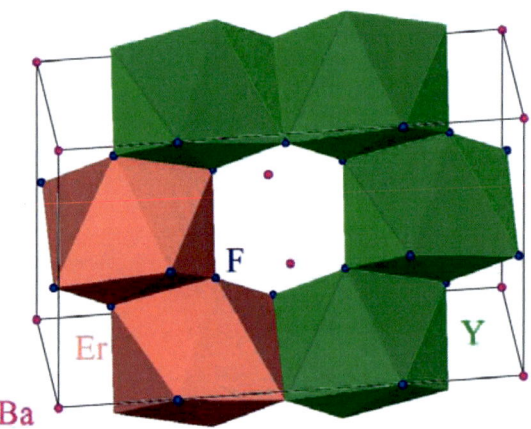

Figure 21. BaY_2F_8 unit cell in which the F^- polyhedra are put in evidence. In the pink octahedra the central Y^{3+} is substituted by Er^{3+}, to show a possible model for a loosely bound Er^{3+} cluster.

C. Er-O Interaction

Other very weak lines, henceforward defined as β, were detected in BaYF samples doped with 0.5, 2, and 12 at% Er^{3+}. Examples are the lines at 6517.4, 6518.84, 6544.9, 6545.7, 6568, 6569.5, 6570, 6572.5, 6590, and 6594.5 cm^{-1}. They cannot be attributed, as the α (Sec. III B), to Er^{3+}-Er^{3+} interaction because they are absent in the sample with the highest concentration (20 at%). However the β lines occur in the typical range of Er^{3+} transitions, therefore they should be tentatively attributed to Er^{3+} perturbed by some unwanted impurity. Possible candidates are OH⁻ and oxygen. OH⁻ can be reasonably ruled out, since no trace of the related vibrational absorption was detected in the infrared, even in very thick samples (7.2 mm) [10]. On the contrary, traces of oxygen could be left behind in the crystal, if the hydrofluorination of starting materials, as Y_2O_3 and Er_2O_3, was not fully accomplished. O^{2-} might substitute for one of the eight F⁻ surrounding the Er^{3+} replacing an Y^{3+}, see Figure 10 (top), thus causing a rather large change of the crystal field probed by Er^{3+} and, as a consequence, of the related energy levels. The strongest among these lines, peaking at 6545.7 cm^{-1}, shows an amplitude decrease as a function of the temperature similar to that displayed by the A1 line at 6530.2 cm^{-1}, therefore could be associated to the A1 line. This means that the perturbation induced by the unwanted impurity is rather large, as an n.n. O^{2-} might cause. In addition, it should be observed that the β line positions, mentioned above, are impressively close to those reported for Er^{3+} in some oxides, as $KY(WO_4)_2$ and $KEr(WO_4)_2$ at 77 K, i.e. 6516, 6544, 6570, and 6600 cm^{-1}, respectively [1]. If O^{2-} substituting for an F⁻ is indeed responsible for the β lines, the large number of them can be accounted for by the required charge compensation, may be by an F⁻ vacancy, which could bring to a variety of Er^{3+}-O^{2-}-F⁻ vacancy complexes. However this remains a hypothesis which should be verified by measuring the spectra of BaYF: Er^{3+} in which the oxygen contamination is intentionally favoured.

D. HYPERFINE STRUCTURE OF Ho^{3+} LINES

Within some RE^{3+} energy level scheme still finer details, as the hyperfine splitting, can be revealed thanks to the high resolution spectroscopy. The topics gained renewed interest, as several rare-earth doped compounds with strong hyperfine interactions have been proposed as possible quantum manipulation media [53,54] after the discovery of simultaneous tunneling of the electronic and nuclear momenta in Ho^{3+}-doped $LiYF_4$ [55]. The hyperfine interaction depends both on the nuclear spin I and on the electronic total angular momentum J [56] being expressed as

$$H_{hf} = A_J \mathbf{J} \cdot \mathbf{I} \qquad (13)$$

within a single manifold, A_J being the related hyperfine coupling constant.

The ^{165}Ho^{3+} ion (natural abundance 100%) has the largest I = 7/2, J = 8 (in the ground state, see Sec. II B 2) and one of the largest A_J (i.e. 0.0271 cm^{-1}) among all the trivalent rare earths.

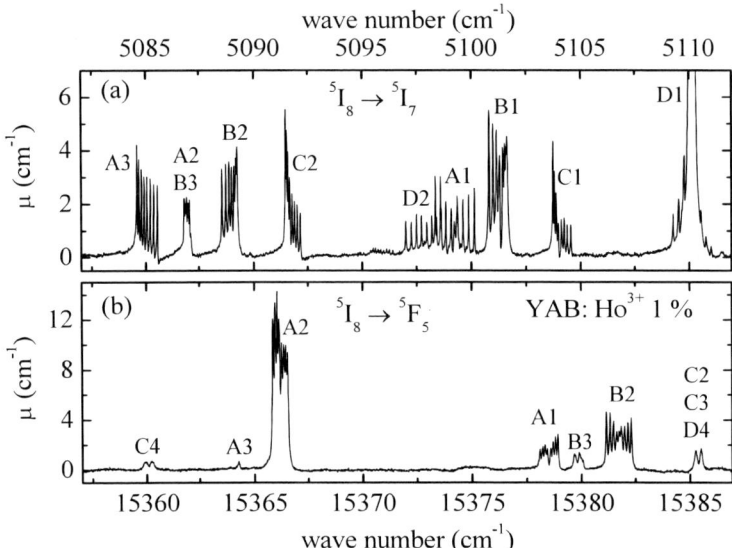

Figure 22. Optical absorption spectra measured at 9 K on a YAB: Ho^{3+} 1 mol% sample in different regions where the Ho^{3+} transitions occur. Panel (a): $^5I_8 \rightarrow ^5I_7$; panel (b): $^5I_8 \rightarrow ^5F_5$. The zero-phonon lines are labeled with Xi (X=A, B,... and i=1, 2,...). Both spectra were measured at 0.01 cm^{-1} resolution.

The first-order approximation simply describes an effective magnetic field A_J <J> which acts on the nuclear spin *I* and produces a Zeeman-like splitting. However, low-symmetry crystal fields (as, for example, those in monoclinic BaYF or rhombic Bi$_2$TeO$_5$ crystals) may lift the manifold degeneracy completely. For non-degenerate states the hyperfine splitting (hfs) is forbidden in the first-order approximation by time reversal symmetry, and thus is not observed [57,58]. On the contrary in YAB, the D$_3$ crystal field experienced by Ho^{3+} cannot remove completely the sublevel degeneracy, thus hyperfine splitting may be observed. By using non-apodized resolution as fine as 0.01 cm^{-1} and by performing the measurements at low temperature (9 K) it was possible to monitor many beautiful hfs patterns affecting the YAB: Ho^{3+} spectra [25,39]. Two examples related to two different transitions ($^5I_8 \rightarrow ^5I_7$ and $^5I_8 \rightarrow ^5F_5$) are portrayed in Figures 22 and 23, respectively. Many finely spaced lines are observed in both spectra, some of them being as narrow as 0.04 cm^{-1}: the structure at about 5085 cm^{-1}, which spans over 1 cm^{-1}, exhibits eight components, see Figure 22(a), as expected for a hyperfine structure due to Ho^{3+} (2I+1=8). Many different hfs patterns were monitored [39]: two examples with eight and nine hfs components are displayed in Figure 23(a) and (b), respectively. Moreover, the hfs components may be either evenly spaced, see the structure at about 5085 cm^{-1} in Figure 22(a), or not, see Figure 23(a). A gap may separate in two groups the eight components, see Figure 23(a). In addition, within a given hfs, the eight components may be either of equal, or of different amplitude, see Figure 23(a).

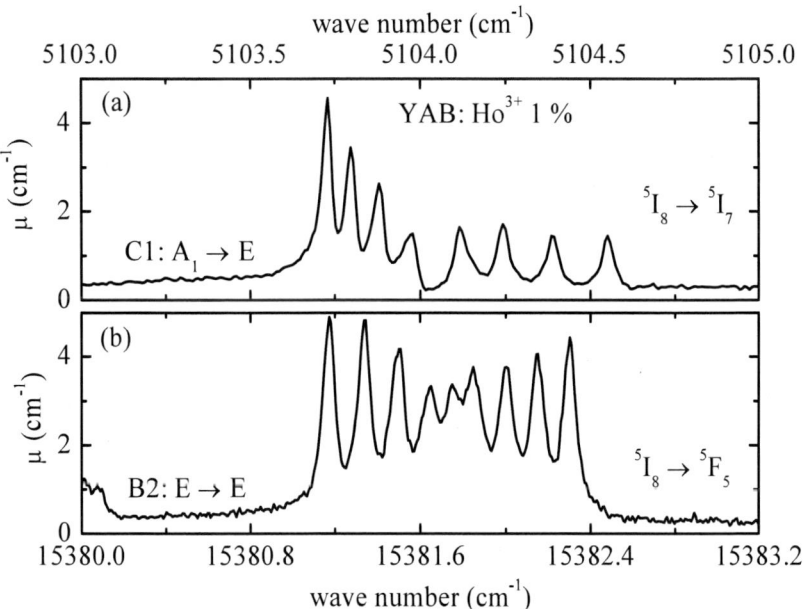

Figure 23. Magnification of optical absorption spectra measured at 9 K on a YAB: Ho^{3+} 1 mol% sample in different regions where the Ho^{3+} transitions occur. Panel (a): C1 line (A$_1$→E) at about 5104 cm^{-1} within the Ho^{3+} ^{5}I$_8$→^{5}I$_7$ transition; panel (b): B2 line (E→E) at about 15381 cm^{-1} within the Ho^{3+} ^{5}I$_8$→^{5}F$_5$. Both spectra were measured at 0.01 cm^{-1} resolution.

Some lines are not split, although remaining narrow, see for example the A3 line at 15364.25 cm^{-1} in Figure 22(b), while in other cases the hyperfine components are broad and merge in a unique line. Due to temperature induced homogeneous broadening the hfs cannot be detected yet at temperatures as low as 40 K.

Necessary preliminary steps to explain the variety of observed hfs patterns are 1) the bare line attribution (i.e. independently of the hfs), by following the procedure reported above (Sections II B 1, II B 2, and II C) and 2) to identify the symmetry of each sublevel state. In the specific case of Ho^{3+} in YAB the states are only either singlets (A_1- and A_2-type) or doublets (E-type) [39], see Sec. II C. In spite of the spectra complexity (Section II B 2) it was possible to provide a convincing interpretation of the features displayed by different hfs structures over a large number of Ho^{3+} transitions ($^5I_8 \rightarrow {}^5I_7$, $^5I_6$, $^5I_5$, $^5F_5$, $^5S_2$, $^5F_4$, $^5F_3$, and $^3K_8$) [39]. The narrow A3 line at 15364.25 cm^{-1} in Figure 22(b) does not exhibit hyperfine splitting for it is originated by a transition between two singlets ($A_1 \rightarrow A_2$). Electric dipole allowed transitions, according to the selection rule $\Delta I_z = 0$ (Figure 24), between a singlet and a doublet (or viceversa) give rise to the eight line hfs, e.g. the C1 line at 5104 cm^{-1} (E$\rightarrow A_1$) in Figure 23(a) and the A3 line at about 5085 cm^{-1} ($A_1 \rightarrow$E) in Figure 22(a). The nine line spectra originate from transitions between two doublets, e.g. the B2 line at 15382 cm^{-1} (E$\rightarrow$E): the eight lines due to electric dipole contribution are accompanied by an additional one which originates from the superposition of the eight lines due to magnetic dipole allowed transitions ($\Delta I_z = 0$, Figure 24) in the specific case where the overall hyperfine splittings of the initial (Δ_i) and final (Δ_f) electronic levels are practically the same [39], see Figure 23(b).

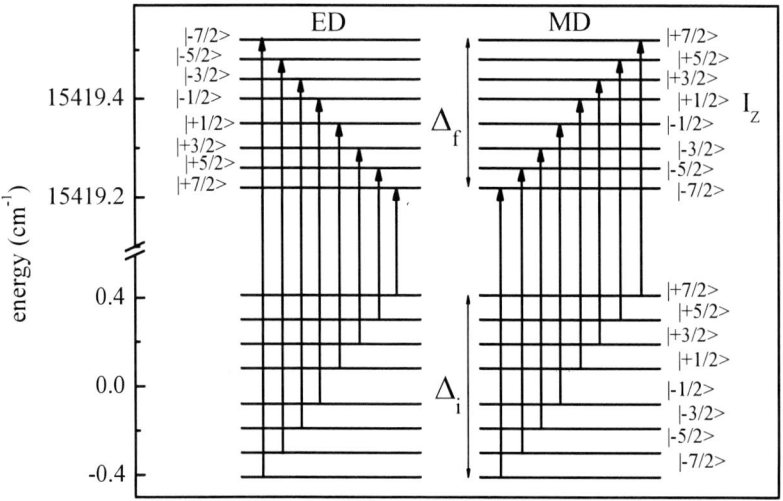

Figure 24. Scheme of the electric dipole (ED) and magnetic dipole (MD) allowed transitions between two multiplets split by the hyperfine interaction in YAB: Ho^{3+}.

The hfs exact calculation can be performed by diagonalizing the full Hamiltonian representing a given manifold in the product space between nuclear and electronic states. The Δ_i and Δ_f values for many sublevels are found to be in excellent agreement with those measured experimentally. Second order effects and slight distortion of Ho^{3+} occupied sites accounted for both the non equispaced hfs patterns and the gap between the two groups of four lines, as displayed by Figure 23(a) [39].

The Ho^{3+} hyperfine structure has been observed by other Authors in different crystals, as $LiYF_4$, CaF_2, and $CsCdBr_3$ [57,59-62], however its detection is seemingly restricted to a very few transitions, as $^5I_8 \rightarrow {}^5I_7$, $^5I_6$, and $^5F_5$. The present results show how the high resolution spectroscopy extended over a wide wave number range provides a comprehensive and detailed description even of very weak RE^{3+} interactions, as that between the 4f electrons and the nuclear momentum.

Chapter IV

ELECTRON-PHONON (E-P) INTERACTION

The analysis of RE^{3+} spectra at different temperatures is not only one of the necessary tools to make a correct line attribution (Secs. II B and II C), but it also supplies valuable insight into the interaction between 4f electrons and the lattice phonons. The study is performed by monitoring the line positions and half maximum full width (HMFW) as a function of the temperature, see Sec. IV A. The related plots are fitted according to the current models to obtain either the coupling constants or the coupled phonon, see Sec. IV B. The e-p interaction is also revealed by the presence of the vibronic tails, i.e. weak structures, which appear on the high energy side of the zero phonon (ZP) lines, i.e. those due to purely electronic transitions. In this case the photon simultaneously excites an electronic transition and a phonon mode, see Secs. IV C and IV D. The vibronic tails in laser materials, as BaYF and YAB, are relevant in view of tuning and extending the laser emission to energies below that of the sharp ZP emission lines.

A. LINE POSITION AND WIDTH AS A FUNCTION OF THE TEMPERATURE

By increasing the temperature the lines, related to RE^{3+} embedded in ordered environments (e.g. crystals and nanocrystals), broaden. Examples are displayed by the comparison of curves a-f in Figures 1 and 2 for Er^{3+} embedded in a variety of crystalline matrices and directly by Figures 4(a), 7(b), and 9(a) for Er^{3+}, Tm^{3+}, and Ce^{3+} doped BaYF crystals, respectively, and by Figure 18 for Er^{3+} in fluorinated nanostructured silica (Secs. III A 2 and III A 3). In addition, the lines shift by increasing the temperature: Figure 7(b), where a magnification of the line peaking at 5683.5 cm^{-1} (i.e. the A1 line within the Tm^{3+} $^3H_6 \rightarrow ^3F_4$ transition in BaYF) is portrayed, clearly shows such a trend. In disordered systems as glasses, the inhomogeneous broadening prevails with respect to the homogeneous one, thus the line width is only poorly affected by the temperature, compare curves i in Figures 1 and 2.

To perform a correct analysis the following criteria are adopted: 1) the spectra are measured at 9 K and then at temperatures increasing from 20 to 300 K by 20 K steps, see Figure 7; 2) only isolated lines are considered (for example, it is not simple to analyze the temperature dependence of X1 and X2 lines in YAB: Dy or in BaYF: Ho where the two lines are separated by only 3.3 and 0.6 cm^{-1}, respectively, see Sec. II B 2); 3) the peak absorbance must be well below 1, to avoid problems due to possible detector non linearity [10], which affects the correct amplitude evaluation and, as consequence, also that of the HMFW; 4) Lorentzian-shaped lines are preferably analyzed. According to criteria 2-4 samples with a moderate RE^{3+} content are selected, to avoid the additional line overlapping [Figures 3, 8, 15(a), and Sec. III B] and a dominant inhomogeneous broadening, see Sec. III A 1. At 9 K in good quality crystals with low dopant content the homogeneous broadening, proved by the line Lorentzian shape, prevails over the inhomogeneous one suggested by Gaussian or Voigt profiles [44]. It was possible to verify that such lines maintained the Lorentzian shape at least up to 140-160 K, see for example Figure 6 in Ref. [10]: at higher temperature the line shape analysis is not easy and may be not reliable due to line weakness and broadening, see for example Figure 7(b).

Figure 25 summarizes the behaviour of the position and width as a function of the temperature for A1 and A2 lines in Er^{3+} doped YAB and BaYF, respectively, and for Ce^{3+} D1 line in BaYF. The lines broaden, see panels (b), (d), and (f), while the line position may show different trends, i.e. a shift either to the higher wave numbers [*blue shift*, see panels (a), (c), and (e) in Figure 25] or to the lower ones (*red shift*, see, for example, the A1 line at 5683.5 cm^{-1} in Tm^{3+} doped BaYF displayed in Figure 7(b)]. The data spreading at relatively high temperatures [see panels (a) and (b) in Figure 25] is due to line broadening and weakening [see, for example, the 'high' temperature curves in Figure 7(b)] and to the consequent difficulty encountered in identifying the maximum position and in drawing correctly the baseline to subtract.

B. MODELS, FITTING, AND ELECTRON-PHONON INTERACTION

Different mechanisms may be responsible for the line broadening in crystals: non-radiative decay or Raman relaxation of the ion to lower energy states involving one or more phonons, resonant or Raman excitation to higher energy states, and Raman scattering of phonons which does not change the ion electronic state. The first four processes affect the line width by shortening the lifetime, while the last (two phonon Raman scattering) broadens the energy levels without removing the electrons. Since the width obtained from the inverse of the decay times is much lower than that measured from the spectra, the two phonon Raman scattering (TPRS) is the main responsible for line broadening [63]. Thus, the temperature dependence of the line width ΔE_i for the ith line can be written as

$$\Delta E_i = \alpha_i \left(\frac{T}{T_D}\right)^7 \int_0^{T/T_D} \frac{x^6 e^x}{(e^x-1)^2} dx \qquad (14)$$

and that of the line shift δE_i as

$$\delta E_i = \beta_i \left(\frac{T}{T_D}\right)^4 \int_0^{T_D/T} \frac{x^3}{e^x-1} dx \qquad (15)$$

where T_D is the Debye temperature, while α_i and β_i are the electron-phonon (e-p) coupling parameters. δE_i and ΔE_i are the line shift and broadening with respect to the position and width at T=0. However, due to the flat behavior of both line position and width at very low temperatures (Figure 25) the shift and broadening were calculated with respect to the values measured at 9 K, respectively. As a rule, T_D, α_i, and β_i are obtained by fitting the experimental data to Eqs. 14 and 15.

Such an approach was followed by Ellens et al. [64,65] to study the e-p interaction. They analyzed the line broadening ΔE_i for a considerable number of RE^{3+} (Pr, Er, Tm, Nd, Eu, Gd, Tb, and Yb) in LiYF$_4$ single crystals and microcrystalline powders, mainly by measuring, at 0.2 cm^{-1} resolution, the fluorescence spectra related to transitions lying at energies higher than 10000 cm^{-1}. The Ce transitions in LiYF$_4$, occurring at lower energies (2000-3500 cm^{-1}), were investigated by means of FT absorption spectroscopy at a resolution of 2 cm^{-1}. They found that the broadening of the high energy lying fluorescence lines could be satisfactorily fitted by Eq. 14, if the Debye temperature was kept as a fixed parameter (T_D was not directly measured) for all the transitions investigated (i.e. T_D=250 K). From the analysis, they could evaluate the coupling constant α for many transitions. A plot of α vs. the number of 4f electrons shows an intriguing trend, i.e. the coupling parameter is high at both sides of the lanthanide series and decreases smoothly towards the middle. The result was explained by considering the superposition of two opposite effects: 1) the RE^{3+} ionic radius contraction (which decreases the interaction of the 4f electrons with the environment) and 2) the weaker and weaker screening operated by the outermost shells on the 4f electrons along the lanthanide series, which favors the interaction [65].

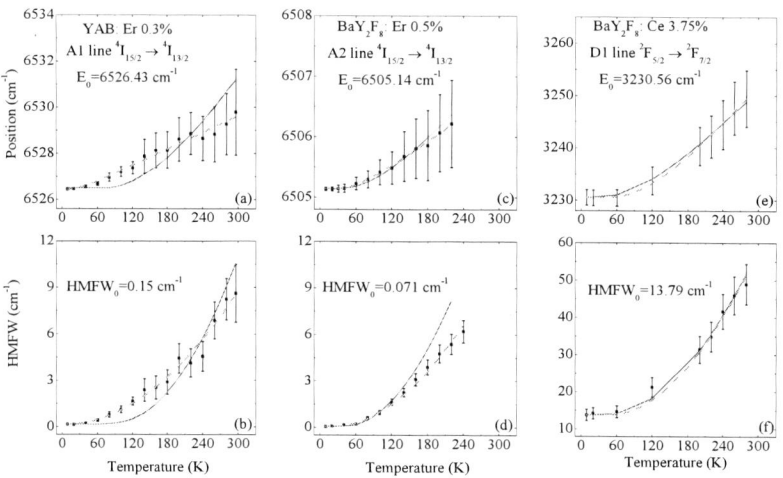

Figure 25. Position and half maximum full width (HMFW) temperature dependence for a few RE3+ absorption lines in different crystals. Black squares: experimental values; red solid line: fitting according to the two phonon Raman scattering model; green dashed line: single phonon fitting. Panels (a) and (b): position and HMFW vs. temperature of A1 line (6526.43 cm-1) within the Er3+ 4I15/2→4I13/2 transition for a YAB: Er3+ 0.3 mol% sample. Panels (c) and (d): position and HMFW vs. temperature of A2 line (6505.14 cm-1) within the Er3+ 4I15/2→4I13/2 transition for a BaY2F8: Er3+ 0.5 at% sample. Panels (e) and (f): position and HMFW vs. temperature of D1 line (3230.56 cm-1) within the Ce3+ 2F5/2→2F7/2 transition for a BaY2F8: Ce3+ 3.75 at% sample. The low temperature spectra were measured at 0.04, 0.02, and 0.1 cm-1 for YAB: Er3+, BaY2F8: Er3+, and BaY2F8: Ce3+ samples, respectively. E0 and HMFW0 are the line position and width extrapolated at T=0.

In the present case the Debye temperature of the host crystal was known, i.e. ~820 and 389 K for YAB and BaYF, respectively, as directly evaluated by means of specific heat measurements in the temperature range 1.5-25 K [66]. Thus it was possible to apply a simple one-parameter fitting to the temperature dependence of ΔE_i and δE_i according to Eqs. 14 and 15, respectively and estimate both α_i and β_i. At variance with previous works [64,65] even lines lying at energies lower than 10000 cm^{-1} were considered. As reported in Sec. III A, such lines are as a rule sharper than those occurring at higher wave numbers and can be easily resolved by exploiting the available high resolution spectrometer (Sec. II A). Narrow lines, if correctly resolved, may work as very sensitive tools to test the models proposed for the line broadening and shift. The experimental data related to the line shift and broadening of many lines due to different RE^{3+} in BaYF and YAB were fitted according to Eqs. 15 and 14, respectively. Examples are shown in Figure 25, where the red curves portray the fitting by following the TPRS model. The fitting result is convincing for the rather broad D1 line in Ce^{3+} (3.75 at%) doped BaYF, see panels (e) and (f) in Figure 25. However, for a few lines in Er^{3+} (0.3 mol%) doped YAB both fittings of line broadening and shift according to Eqs. 14 and 15 fail, as shown for the A1 line at 6526.4 cm^{-1} [see panels (b) and (a), respectively]. An intermediate situation is displayed by the A2 line in Er^{3+} (0.5 at%) doped BaYF, where the fitting according to the TPRS model appears successful for the line shift, see panel (c), but not for the line broadening, see panel (d). Therefore a different approach was followed, by applying the single phonon coupling (SPC) model [67], which was already employed successfully to analyze line shift and broadening of OH stretching mode absorption in alkali halides, KMgF$_3$, and KTP [45,68,69] and of two-photon emission lines in Eu^{2+} doped KMgF$_3$ [70].

In the framework of the SPC model δE_i and ΔE_i are expressed by

$$\delta E_i = \delta \omega \left[\exp\left(\frac{\hbar \omega_0}{k_B T}\right) - 1 \right]^{-1} \tag{16}$$

$$\Delta E_i = \frac{2(\delta \omega)^2}{\gamma} \exp\left(\frac{\hbar \omega_0}{k_B T}\right) \left[\exp\left(\frac{\hbar \omega_0}{k_B T}\right) - 1 \right]^{-2} \tag{17}$$

where $\hbar \omega_0$ and γ are energy and width of the coupled phonon band, $\delta \omega$ is the coupling constant, which is assumed to be smaller than γ, and k_B is the Boltzmann constant. The experimental data, displayed in Figure 25 and related to line shift and broadening, were fitted according to Eqs. 16 and 17, see green dashed curves. In panels (a)-(d) the fittings according to the SPC model are excellent, while in panels (e) and (f) no meaningful difference may be detected between the green and red curves, related to the SPC and TPRS models, respectively.

A parameter with an important physical meaning can be derived from the fitting according to SPC model, i.e. the coupled phonon energy $h\omega_0$, which may be compared with the values provided by IR absorption and/or Raman spectroscopy. For example in the case of A1 line at 6526.4 cm^{-1} related to Er^{3+} in YAB, both the fittings according to Eqs. 16 and 17 [green dashed curves in panels (a) and (b), respectively] were found to supply, independently, the same value for the energy of the coupled phonon ($h\omega_0 \sim 50$ cm^{-1}). It is worthwhile to remark that such a value practically coincides with a Raman active mode (51 cm^{-1}) of the host matrix [66], see Table 9. Furthermore, the same phonon was found to replicate the ZP lines in the vibronic tails of YAB: Dy^{3+}, see Table 9 and Sec. IV D. Similar results were obtained for other lines induced by the Er^{3+} $^4I_{15/2} \rightarrow {}^4I_{13/2}$ transition in YAB: the coupled phonons, deduced from the fittings according to the SPC model are found to be in nice agreement either with IR and/or Raman active vibrational modes, see Table 9. The question now is why for some lines the experimental data can be successfully fitted only in the framework of the SPC model. Inspection of Figure 25 provides some hints. The shift and broadening of the A1 line at 6526.4 cm^{-1} in Er^{3+} doped YAB according to the SPC model is caused by the interaction with a single phonon which lies at much lower energy (~ 50 cm^{-1}) than the upper limit (~ 1400 cm^{-1} [71]) of the host crystal phonon spectrum. On the contrary, in the TPRS model the whole phonon spectrum is considered in the e-p interaction as it appears clearly from Eqs. 14 and 15, where the integration over the temperature is extended up to the Debye temperature. At variance, for the Ce^{3+} D1 line at 3230.6 cm^{-1} in BaYF, where the fitting according to TPRS and SPC models are both satisfactory [panels (e) and (f)], the phonon, obtained from the SPC model, is ~ 240 cm^{-1}, i.e. not too far from the upper limit (~ 450 cm^{-1} [72]) of BaYF phonon spectrum. An important role is played also by the line width at low temperature (HMFW$_0$), in fact, by considering the same host, e.g. BaYF, the line broadening is well fitted even by the TPRS model in the case of the broad Ce^{3+} D1 line (HMFW$_0 \sim 13.8$ cm^{-1}), but not for the narrow Er^{3+} A2 line (HMFW$_0 \sim 0.07$ cm^{-1}), compare panels (d) and (f) in Figure 25. The RE^{3+} concentration is also expected to affect the ΔE_i temperature dependence for a given line, because inhomogeneous broadening occurs at low temperature for high RE^{3+} doping levels, see Sec. III A 1: by increasing the Er concentration in BaYF from 0.5 at% to 2 at%, the departure, shown in panel (d), of the experimental data from the TPRS model for the 0.5 at% Er doped sample (red curve) was significantly attenuated for the 2 at% doped one. Moreover, if not high enough instrumental resolution is used to analyze the temperature induced broadening of narrow lines, the measured HMFW remains practically constant vs. temperature until the line is broad enough to be correctly resolved. As a consequence, the experimental data of HMFW plot vs. temperature mime the flat trend exhibited at low temperatures by the TPRS fitting, see for example red curves in panels (b) and (d). Therefore reliable conclusions on the model which better describes the e-p interaction in a given crystalline matrix can be drawn by using diluted RE^{3+} solid solutions and by measuring the optical spectra with a resolution high enough to resolve properly even the narrow lines.

Table 9. Comparison between phonon energies evaluated from the fitting according to the SPC model for A1, B1, C1, and D1 lines within the $^4I_{15/2} \rightarrow {}^4I_{13/2}$ transition of Er^{3+} in YAB (last column) with those obtained from vibrational IR and Raman spectra, respectively

Line	Position (cm^{-1})	Raman (RT) X(ZZ)X (cm^{-1})	Raman (RT) Z(XX)Z (cm^{-1})	IR (9 K) (cm^{-1})	$\Delta\omega_{ZP}$ (9 K) (cm^{-1})	SPC fitting (cm^{-1})
A1	6526.43 ±0.13	51			50.03±1.63	51.6
B1	6560.78±0.09	110			110.05±0.89	110.3
C1	6611.3 ±0.05	118.7	118.3	116.4	118.3±1.9	114
D1	6639.55 ±0.06		135	~132	134.4±1.2	132.1

The vibronic shift $\Delta\omega_{ZP}$ by the same phonons observed in YAB: Dy^{3+} are reported in the last but one column.

C. VIBRONIC TRANSITIONS

An absorption vibronic transition takes place between the states $\psi(a, n)$ and $\psi(b, n')$, where the final vibrational state n' differs from the initial one n, a and b indicate the ground and the final electronic states, respectively [73]. The optical transition from a to b consists of a number of vibronic lines, which are characterized by the vibrational states involved, n and n'. At low temperatures only the vibrational level $n=0$ is occupied in the ground electronic state a. The transition to $n'=0$ is called zero-phonon (ZP) transition: this originates a very sharp line. Sections II, III, IV A, and IV B are completely devoted to the ZP line spectra. The vibronic transitions 0-n' appear in the absorption spectra at higher energies with respect to the ZP line (vibronic replica). By using the Condon approximation the matrix element for the transition is given by $<\psi(b)|\mu|\psi(a)><\chi(n')|\chi(n)>$ where χ denotes the vibrational wavefunction and μ is the electric dipole operator. If intraconfigurational 4f-4f transitions are considered (as in the case of RE^{3+}), the electronic matrix element, i.e. $<\psi(b)|\mu|\psi(a)>$ vanishes for electric dipole transitions. For $n=0$, $|\chi(n)>$ has the totally symmetric representation, therefore the above matrix element does not vanish only if $|\chi(n')>$ has the same representation as the electric dipole operator μ. Thus, parity forbidden electric dipole transitions are in part allowed by interaction with ungerade vibrational modes. The selection rule for the vibronic transitions accompanying the ZP transitions of RE^{3+} is $\Delta J \leq 2$, where ΔJ is the difference between the total quantum numbers of the ground and final electronic states.

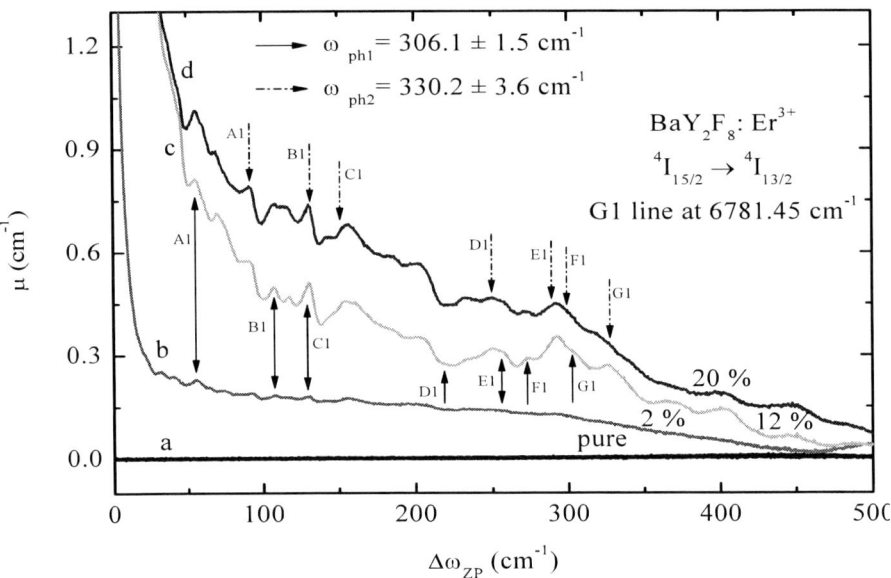

Figure 26. Vibronic tails in the region of Er^{3+} $^4I_{15/2} \rightarrow ^4I_{13/2}$ transition for BaY_2F_8 samples doped with different Er^{3+} concentrations as a function of the reduced wave number $\Delta\omega_{ph}$. All spectra are measured at 9 K. Curve a: pure sample; curve b: 2 at%; curve c: 12 at%; curve d: 20 at%. The arrows indicate the vibronic replica of the seven ZP lines at energies higher than 306.1 (solid arrows) and 330.2 cm^{-1} (dot-dashed arrows), respectively.

D. VIBRONIC SPECTRA ANALYSIS

In Figure 9(a) a fine structure appears on the high wave number side of the A1 line (~2197.5 cm^{-1}) and overlaps to B1 and C1 lines of a BaYF sample doped with 3.75 at% Ce: this structure is present in spectra measured at 9 and 60 K (curves a and b) and vanishes at higher temperatures, see curves c and d (at 120 and 300 K, respectively). By analyzing the high energy side of the last Xn line (within a given transition between the ground and an excited manifold, see Sec. II B 1), similar fine structures were put in evidence for a variety of RE^{3+}, transitions, and crystalline hosts (e.g. BaYF and YAB). The amplitude of the fine structure component lines is much weaker than that of the lines originated by the 4f-4f electronic transitions (ZP lines), compare, for example, the ordinate scale of Figure 27(b) (for the fine structure) with that of Figure 27(a), which displays the ZP spectrum originated by the $^3H_6 \rightarrow {^3F_3}$ transition of Tm (5 at%) in BaYF. The weak lines are closely related to the presence of the RE^{3+}, as clearly shown in Figure 26 for BaYF doped with different amounts of Er^{3+}: the lines, which lye on the high energy tail of the G1 line within the $^4I_{15/2} \rightarrow {^4I_{13/2}}$ transition, are absent in the pure sample (curve a) and their amplitude increases by increasing the Er^{3+} concentration (curves b-d). Moreover the fine structure is peculiar of a given RE^{3+}, e.g. Tm^{3+}, as shown in Figure 27(b), where curves a and b portray the high energy tail of Tm^{3+} G1 line in a Tm^{3+} and in a Tm^{3+}-Ho^{3+} doped BaYF sample, respectively: there is no difference between the two spectra.

The weak lines are attributed to vibronic transitions, i.e. the incident photon simultaneously excites an electronic transition and a phonon mode [71,74], see Sec. IV C. Since two processes are involved in vibronic transitions, the related spectra are much weaker than those related to ZP transition, as observed above and proved by Figures 27(a) and (b). The question is how to identify, within the lattice vibrations, the specific phonons which take part in the vibronic transitions. The task is not easy, because the phonon spectra extension is ~450 [72] and ~1400 cm^{-1} [71] for BaYF and YAB, respectively. This means that the vibronic replica of every Xn line may cover the above mentioned wave number ranges on every ZP line high energy side. Thus, different replicas may mix one to each other giving rise to very crowded vibronic tails, as shown, for example, in Figure 26 for BaYF: Er^{3+}. The guidelines which allowed accomplishing the purpose are the following:

1) to acquire information about the lattice modes by measuring the IR spectra in the region of lattice absorption or by means of Raman scattering measurements;
2) to measure the wave number 'separations' $\Delta\omega_{ZP}$ of the vibronic peaks from the corresponding ZP lines to evaluate the energy of phonons involved in the vibronic transition: an example of vibronic spectra vs. reduced wave number (i.e. by subtracting from the actual wave number that of a given Xn line) is displayed in Figure 26 for BaYF crystals doped with different amounts of Er^{3+};
3) to compare the values so obtained with the IR and/or Raman-active vibrational modes determined according to 1).

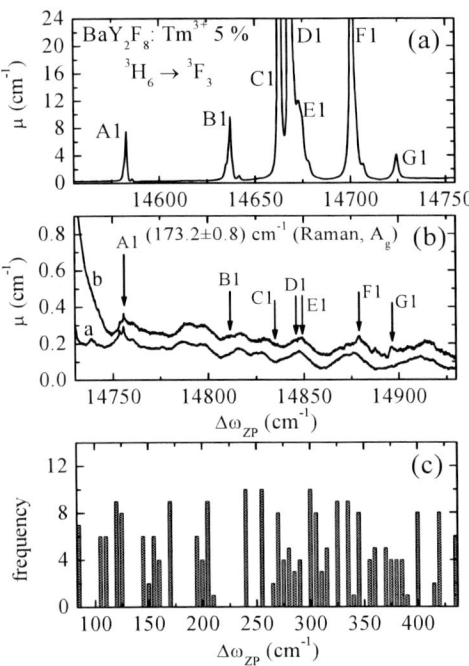

Figure 27. Panel (a): optical absorption spectrum measured at 9 K in the region of Tm^{3+} $^3H_6 \rightarrow {}^3F_3$ transition for a BaY_2F_8: Tm^{3+} 5 at% sample. The ZP lines are labelled with X1 (X=A, B,...). Panel (b): vibronic tails related to the Tm^{3+} $^3H_6 \rightarrow {}^3F_3$ transition for BaY_2F_8 samples. Curve a: BaY_2F_8: Tm 5 at%; curve b: BaY_2F_8: Tm^{3+} 5.2 at%, Ho^{3+} 0.5 at%. The solid arrows indicate the vibronic replica of the ZP lines by the phonon at 173.2 cm^{-1}. Panel (c): histogram of the phonons involved in the vibronic replica of BaY_2F_8: Tm^{3+} 5 at% in the region of Tm^{3+} $^3H_6 \rightarrow {}^3F_3$ transition.

To fulfill the first requirement the absorption spectra of thin CsI pellets, in which BaYF or YAB powders had been diluted (Sec. II A), were measured at 9 K in the wave number ranges 75-500 and 75-1400 cm^{-1}, respectively. In this way the IR active vibrational modes were identified for both matrices [71,74]. The Raman active modes were evaluated by measuring the Raman spectra at RT in the range 7-1500 [66]. By following the second guideline a plenty of $\Delta\omega_{ZP}$ values, i.e. of possible lattice vibrational modes ω_{ph}, were collected, due to the large number of transitions investigated for different RE^{3+} in the same host crystal. For example, in the case of BaYF $\Delta\omega_{ZP}$ could be evaluated for Ce^{3+}, Dy^{3+}, Er^{3+}, and Tm^{3+} over a total number of 19 inter-manifold transitions: in addition for each of them many ZP lines were considered. The frequency by which each $\Delta\omega_{ZP}$ appears in the vibronic tails accompanying the ZP lines of a given transition is plotted in a histogram as that displayed by Figure 27(c) for the Tm^{3+} $^3H_6 \rightarrow {}^3F_3$ transition in BaYF. Clearly many $\Delta\omega_{ZP}$ values stand out well above a noise level, predicted for a random distribution. Still clearer cut evidence for some $\Delta\omega_{ZP}$ is obtained, if the $\Delta\omega_{ZP}$, calculated for all dopants and transitions, are collected in a unique histogram for a given crystal, due to the very large amount of acquired data. However, the conclusive proof that the outstanding values in the histograms are indeed the lattice modes ω_{ph} involved in the vibronic transitions comes from their comparison with those monitored by IR and Raman spectra measured on the host matrix sample. A large number of coincidences were found to support the attribution of the weak lines to vibronic transitions for different RE^{3+} both in BaYF and YAB. In many cases, it was possible to identify within the vibronic tails the replica (by a given phonon ω_{ph}) of the entire sequence of Xn lines originated by a 4f-4f transition. Two examples are portrayed in Figure 26 for Er^{3+} in BaYF, where the whole set of A1-G1 lines, originated by the $^4I_{15/2} \rightarrow {}^4I_{13/2}$ transition, is replicated by the weaker vibronic tail, which is shifted toward higher wave numbers by two vibrational quanta, i.e. ~306 and 330 cm^{-1}, respectively. The two values coincide with phonons which are both IR and Raman active. In principle only IR-active phonons are expected to couple more effectively to the electronic parity forbidden 4f-4f transitions (Sec. IV C), however a few modes are simultaneously IR- and Raman-active in BaYF, supporting a slightly disordered structure of the host crystal, as also suggested by x-ray diffraction measurements [51] and crystal field calculations [10,22]. In Figure 27(a) the ZP spectrum originated by the $^3H_6 \rightarrow {}^3F_3$ transition of Tm^{3+} in BaYF is displayed just over its vibronic replica [Figure 27(b)] shifted to higher wave numbers by ω_{ph}~173 cm^{-1} (an A_g Raman active phonon), to provide a still more convincing comparison. A further support to the attribution of the weak structures to transitions, in which a host matrix phonon is involved, is given by Figure 28 where the vibronic tails are compared for different RE^{3+} (and obviously different transitions) in the same host (BaYF): the arrows indicate how some vibronic peaks are shifted by the same phonon with respect to the considered ZP line.

The selection rule $\Delta J \leq 2$ (Sec. IV C) was observed for most of the vibronic replicas, see the example of Figure 26 for the Er^{3+} $^4I_{15/2} \rightarrow ^4I_{13/2}$ transition, however exceptions were also detected, see the example supplied by Figure 27 where the vibronic replica for the Tm^{3+} $^3H_6 \rightarrow ^3F_3$ transition is portrayed. Violations of the selection rule are reported for other RE^{3+} in BaYF and for other systems [74, 75]: in the present case it can be accounted for by the low symmetry (C_2) of the RE^{3+} site in BaYF, see Sec. II C and Figure 10.

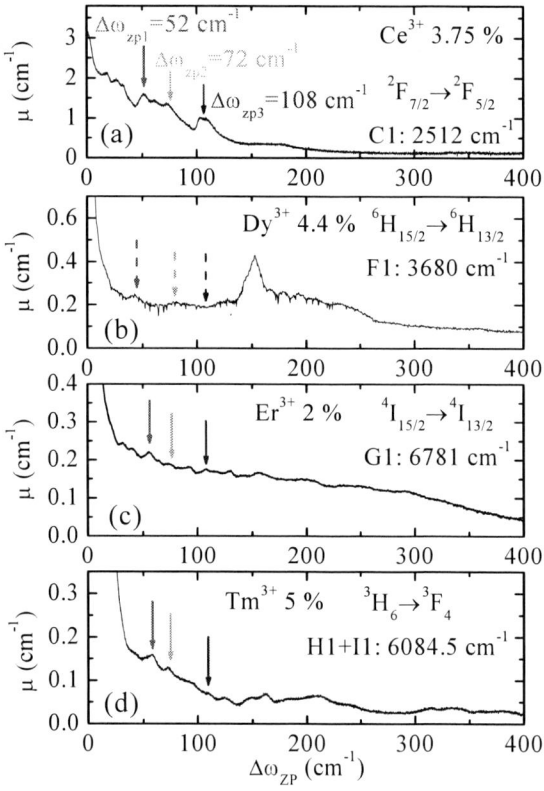

Figure 28. Comparison between the BaY_2F_8 vibronic tails of different ions. Panel (a): BaY_2F_8: Ce^{3+} 3.75 at%. Panel (b): BaY_2F_8: Dy^{3+} 4.4 at%. Panel (c): BaY_2F_8: Er^{3+} 2 at%. Panel (d): BaY_2F_8: Tm^{3+} 5 at%. The arrows indicate the vibronic lines shifted by 52, 72, and 108 cm^{-1} phonons with respect to the ZP lines, respectively.

Chapter V

CONCLUSION

The results obtained by our group in the last years [9-11,14,15,17-20,22-25,27,39,45-47,58,68,69,71,74,76] clearly prove how the FT absorption spectroscopy, coupled to theoretical calculations, is able to thoroughly exploit the unique feature offered by the nearly atomic character of the optical transitions induced by RE in insulating materials. However necessary preconditions are required, as high resolution (as fine as 0.01 cm^{-1}), wide spectral range covered (75-25000 cm^{-1}), a close step temperature analysis (9-300 K), and different doping levels. According to the literature, even better resolutions were occasionally employed to investigate some subtle effects due to RE^{3+} [43,59], however the analysis was always restricted to a limited spectral range.

In addition to supplying a complete and accurate energy level scheme for a large number of RE and different hosts, the procedure adopted in the present work provided a deep insight on a variety of environments probed by RE (e.g. clusters and nanocrystals in glass-ceramics), and on the 4f electron interactions with both nuclear spins and lattice vibrations.

ACKNOWLEDGMENTS

The Authors like to express their gratitude to Prof. Giuseppe Amoretti (Physics Department, University of Parma) for helpful suggestions and discussions on the theoretical modeling, Prof. Giuseppe Carini and co-workers (Physics Department, University of Messina) for Raman spectra and heat capacity measurements, Prof. Mauro Tonelli and co-workers (Physics Department, University of Pisa) for providing BaY_2F_8 crystals, Dr. István Földvári and co-workers (Research Institute for Solid State Physics and Optics, Budapest, Hungary) for supplying YAB crystals, Dr. Norberto Chiodini, Prof. Alberto Paleari, Prof. Anna Vedda, and co-workers (Department of Materials Science, University of Milano-Bicocca) for providing glasses and glass ceramics samples, Dr. Francesca Licci (IMEM-CNR, Parma) for supplyng a YAG: Er crystal, Carlo Mora (IMEM-CNR, Parma) for technical help. The Authors wish to thank a few undergraduate and graduate students of the Parma University (Andrea Ruffini, Marina Cornelli, Alberto Sperzagni, Andrea Ponzoni, Peter Riolo, Andrea Losi, Vittorio Cannas, and Lorenzo Venturi) for some high resolution measurements. The financial support of Italian MiUR is also acknowledged.

REFERENCES

[1] Kaminskii, A. A. *Laser Crystals;* second edition; Springer-Verlag: Berlin, 1990. Kaminskii, A. A. *Crystalline Lasers: Physical Processes and Operating Schemes;* CRC Press: New York, NY, 1996.
[2] Rodnyi, P. A. *Physical Processes in Inorganic Scintillators;* CRC Press: New York, NY, 1997.
[3] Joubert, M. -F.; Kazanskii, S. A.; Guyot, Y.; Gâcon, J. -C; Pedrini C. *Phys. Rev. B* 2004, *69*, 165217-1-13.
[4] Thiel, C. W.; Cruguel, H.; Wu, H.; Sun, Y.; Lapeyre, G. J.; Cone, R. L.; Equall, R. W.; Macfarlane, R. M. *Phys. Rev. B* 2001, *64*, 085107-1-27.
[5] Kaminskii, A. A. *Laser & Photon. Rev.* 2007, *1*, 93-177.
[6] Dieke, G. H. *Spectra and Energy Levels of Rare Earth Ions in Crystals;* J. Wiley & Sons: New York, NY, 1968.
[7] Toncelli, A.; Tonelli, M.; Cassanho, A.; Jenssen, H. P. *J. Lumin.* 1999, *82*, 291-298.
[8] Beregi, E.; Hartmann, E.; Malicskó, L.; Madarász, J. *Cryst. Res. Techn.* 1999, *34*, 641-645.
[9] Földvári, I.; Beregi, E.; Baraldi, A.; Capelletti, R.; Munoz, A.; Sosa, R.; Ryba-Romanowski, W.; Dominiak-Dzik, G. *J. Lumin.* 2003, *102-103*, 395-401.
[10] Baraldi, A.; Capelletti, R.; Mazzera, M.; Ponzoni, A.; Amoretti, G.; Magnani, N.; Toncelli, A.; Tonelli, M. *Phys. Rev. B* 2005, *72*, 075132-1-16.
[11] Baraldi, A.; Capelletti, R.; Magnani, N.; Mazzera, M.; Beregi, E.; Földvári, I. *J. Phys. C Condensed Matter* 2005, *17*, 6245-6255.
[12] Chiodini, N.; Meinardi, F.; Morazzoni, F.; Padovani, J.; Paleari, A.; Scotti, R.; Spinolo, G. *J. Mater. Chem.* 2001, *11*, 926-929.

[13] Chiodini, N.; Paleari, A.; Di Martino, D.; Spinolo, G. *Appl. Phys. Lett.* 2002, *81*, 1702-1704.
[14] Baraldi, A.; Buffagni, E.; Capelletti, R.; Mazzera, M.; Brovelli, S.; Chiodini, N.; Lauria, A.; Moretti, F.; Paleari, A.; Vedda, A. *J. Non-Cryst. Solids* 2007, *353*, 564-567.
[15] Brovelli, S.; Baraldi, A.; Capelletti, R.; Chiodini, N.; Lauria, A.; Mazzera, M.; Monguzzi, A.; Paleari, A. *Nanotechnology* 2006, *17*, 4031-4036.
[16] Di Martino, D.; Vedda A.; Angella, G.; Catti, M.; Cazzini, E.; Chiodini, N.; Morazzoni, F.; Scotti, R.; Spinolo, G. *Chem. Mater.* 2004, *16*, 3352-3356.
[17] Magnani, N.; Baraldi, A.; Buffagni, E.; Capelletti, R.; Mazzera, M.; Brovelli, S.; Lauria, A. *Phys. Status Solidi (c)* 2007, *4*, 1209-1212.
[18] Földvári, I.; Beregi, E.; Baraldi, A.; Capelletti, R.; Munoz, A.; Sosa, R. *Radiat. Eff. Defect Solids* 2003, *158*, 285-288.
[19] Földvári, I.; Beregi, E.; Capelletti, R.; Baraldi, A. *Phys. Status Solidi (c)* 2005, *2*, 260-263.
[20] Baraldi, A.; Capelletti, R.; Cornelli, M.; Ponzoni, A.; Ruffini, A.; Sperzagni, A.; Tonelli, M. *Radiat. Eff. Def. Sol.* 2001, *155*, 349-353.
[21] Agladze, N. I.; Popova, M. N.; Vinogradov, E. A.; Murina, T.M.; Zhekov, V. I. *Opt. Commun.* 1988, *65*, 351-354.
[22] Magnani, N.; Amoretti, G.; Baraldi, A.; Capelletti, R. *Eur. Phys. J. B*, 2002, *29*, 79-84.
[23] Magnani, N.; Amoretti, G.; Baraldi, A.; Capelletti, R. *Radiat. Eff. Def. Sol.* 2002, *157*, 921-926.
[24] Amoretti, G.; Baraldi, A.; Capelletti, R.; Magnani, N.; Mazzera, M.; Riolo, P.; Sani, E.; Toncelli, A.; Tonelli, M. *Phys. Status Solidi (c)* 2005, *2*, 248-251.
[25] Baraldi, A.; Földvári, I.; Capelletti, R.; Magnani, N.; Mazzera, M.; Beregi, E. *Phys. Status Solidi (c)* 2007, *4*, 1364-1367.
[26] Johnson, L. F.; Guggenheim, H. J. *Appl. Phys. Lett.* 1971, *19*, 44-47.
[27] Baraldi, A.; Capelletti, R.; Mazzera, M.; Amoretti, G.; Magnani, N.; Chiodini, N.; Di Martino, D.; Paleari, A.; Spinolo, G.; Vedda, A. *Phys. Status Solidi (c)* 2005, *2*, 572-575.
[28] Johnson, L. F.; Guggenheim, H. J. *IEEE J. Quantum Electron.* 1974, *10*, 442-449.
[29] Schlafer H. L.; Gliemann, G. *Basic Principles of Ligand Field Theory;* Wiley Interscience: London, 1969.
[30] Mazzera, M. *Fourier Transform Spectroscopy of Rare Earths in Insulating Materials for Photonics;* Ph. D. Thesis, University of Parma, Italy, 2006.

References

[31] Dominiak-Dzik, G.; Solarz, P.; Ryba-Romanowski, W.; Beregi, E.; Kovács, L. *J. Alloys Compounds* 2003, *359*, 51-58.
[32] Wybourne, B. G. *Spectroscopic Properties of Rare Earths;* Interscience: New York, NY, 1965.
[33] Nielson, C. W.; Koster, G. F. *Spectroscopic Coefficients for p^n, d^n and f^n configurations;* MIT Press: Cambridge, MA, 1964.
[34] Judd, B. R. *J. Chem. Phys.* 1966, *44*, 839-840.
[35] Marvin, H. *Phys. Rev.* 1947, *71*, 102-110.
[36] Judd, B. R.; Crosswhite, H. M.; Crosswhite, H. *Phys Rev.* 1968, *169*, 130-138
[37] Carnall, W. T.; Goodman, G. L.; Rajnak, K.; Rana, R. S. *J. Chem. Phys.* 1989, *90*, 3443-3457.
[38] Newman, D. J.; Ng, B. K. C. *Crystal Field Handbook;* Cambridge University Press, 2000.
[39] Baraldi, A.; Capelletti, R.; Mazzera, M.; Magnani, N.; Földvári, I.; Beregi, E. *Phys. Rev. B* 2007, *76*, 165130-1-10.
[40] Hutchings, M. T. In *Solid State Physics;* Seitz, F. and Turnbull, D.; Ed.; Academic Press: New York, NY, 1964; Vol. 3.
[41] Newman, D. J.; Ng, B. *Rep. Prog. Phys.* 1989, *52*, 699-763.
[42] Freeman, A. J.; Desclaux, J. P. *J. Magn. Magn. Mater.* 1979, *12*, 11-21.
[43] Macfarlane, R. M.; Cassanho, A.; Meltzer, R. S. *Phys. Rev. Lett.* 1992, *69*, 542-545.
[44] Svelto, O. *Principle of Lasers;* Plenum Press: New York, NY, 1989; pp 30-42.
[45] Baraldi, A.; Bertoli, P.; Capelletti, R.; Ruffini, A.; Scacco, A. *Phys. Rev. B* 2001, *63*, 134302-1-12.
[46] Baraldi, A.; Capelletti, R.; Chiodini, N.; Mora, C.; Scotti, R.; Uccellini, E.; Vedda, A. *Nucl. Instr. and Meth.* 2002, *486*, 408-411.
[47] Chiodini, N.; Fasoli, M.; Martini, M.; Spinolo, G.; Vedda, A.; Nikl, M.; Solovieva, N.; Baraldi, A.; Capelletti, R. *Appl. Phys. Lett.* 2002, *81*, 4374-4376.
[48] Saito, K.; Ikushima, A. J. *J. Appl. Phys.* 2002, *91*, 4886-4890.
[49] Painter, G. S.; Becher, P. F.; Kleebe, H. -J.; Pezzotti, G. *Phys. Rev. B* 2000, *65*, 064113-1-11.
[50] Agekyan, V. T. *Phys. Status Sol. A* 1977, *43*, 11-42.
[51] Guilbert, L. H.; Gesland, J. Y.; Bulou, A.; Retoux, R. *Mat. Res. Bull.* 1993, *28*, 923-930.
[52] Mészáros, Gy.; Sváb, E.; Beregi, E.; Watterich, A.; Tóth, M. *Physica B* 2000, *276-278*, 310-311.

[53] Bertaina, S.; Gambarelli, S. A.; Tkachuk, I.; Kurkin, N.; Malkin, B.; Stepanov, A.; Barbara, B. *Nat. Nanotechnol.* 2007, *2*, 39-42.
[54] Guillot-Noël, O.; Goldner, Ph.; Antic-Fidancev, E.; Le Gouët, J. L. *Phys. Rev. B* 2005, *71*, 174409-1-15.
[55] Giraud, R.; Wernsdorfer, W.; Tkatchuk, A. M.; Mailly, D.; Barbara, B. *Phys. Rev. Lett.* 2001, *87*, 057203-1-4.
[56] Abragam, A.; Bleaney, B. *Electron Paramagnetic Resonance of Transition Ions;* Clarendon Press, Oxford, 1970.
[57] Agladze, N. I.; Popova, M. N. *Sol. State Commun.* 1985, *55*, 1097-1100.
[58] Földvári, I.; Baraldi, A.; Capelletti, R.; Magnani, N.; Sosa F., R.; Munoz F., A.; Kappers, L. A.; Watterich, A. *Optical Materials* 2007, *29*, 688-696.
[59] Popova, M. N.; Agladze, N. I. *Mol. Phys.* 2004, *102*, 1315-1318.
[60] Martin, J. P. D.; Boonyarith, T.; Manson, N. B.; Mujaji, M.; Jones, G. D. *J. Phys. Condens. Matter* 1993, *5*, 1333-1348.
[61] Strickland, N. M.; Jones, G. D. *Mol. Phys.* 2004, *102*, 1345-1349.
[62] Mujaji, M.; Jones, G. D.; Syme, R. W. G. *Phys. Rev. B* 1993, *48*, 710-725.
[63] Di Bartolo, B. *Optical Interaction in Solids*; John Wiley & Sons: NY, 1968; 341-403.
[64] Ellens, A.; Andres, H.; ter Heerdt, M. L. H.; Wegh, R. T; Meijerink, A.; Blasse, G. *J. Lumin.* 1996, *66-67*, 240-243.
[65] Ellens, A.; Andres, H.; ter Heerdt, M. L. H.; Wegh, R. T.; Meijerink, A.; Blasse, G. *Phys. Rev. B* 1997, *55*, 180-186.
[66] Carini, G. private communication.
[67] Dumas, P.; Chabal, Y. J.; Higashi, G. S.; *Phys. Rev. Lett.* 1990, *65*, 1124-1127.
[68] Beneventi, P.; Bertoli, P.; Capelletti, R. *Mikrochim. Acta Suppl.* 1997, *14*, 491-492.
[69] Baraldi, A.; Capelletti, R.; Kovács, L.; Mora, C. *Radiat. Effects and Defects in Sol.* 2001, *155*, 355-359.
[70] Francini, R.; Grassano, U. M. *Il Nuovo Cimento* 1998, *20D*, 875-883.
[71] Mazzera, M.; Baraldi, A.; Capelletti, R.; Beregi, E.; Földvári, I. *Phys. Status Solidi (c)* 2007, *4*, 860-863.
[72] Kaminskii, A. A.; Sarkisov, S. E.; Below, F. Eichler, H. -J. *Opt. Quantum Electron.* 1990, *22*, S95-S105.
[73] G. Blasse, *Intern. Rev. Phys. Chem.* 1992, *11*, 71-100.
[74] Baraldi, A.; Capelletti, R.; Mazzera, M.; Ponzoni, A.; Sani, E.; Tonelli, M. *Radiat. Eff. Defect Solids* 2003, *158*, 241-245.
[75] de Mello Donegà, C.; Meijerink, A.; Blasse, G. *J. Phys. Condens. Matter* 1992, *4*, 8889-8902.

[76] Baraldi, A.; Capelletti, R.; Magnani, N.; Mazzera, M. *Phys. Status Solidi (c)* 2007, *4*, 778-783.

INDEX

A

absorption, vii, 1, 3, 4, 5, 9, 10, 11, 13, 14, 16, 18, 19, 20, 21, 22, 23, 33, 35, 36, 38, 40, 41, 42, 43, 44, 45, 47, 48, 49, 55, 56, 57, 58, 59, 62, 63, 65
absorption coefficient, 16, 45
absorption spectra, 1, 3, 4, 5, 10, 11, 18, 19, 20, 21, 23, 33, 38, 40, 41, 43, 44, 48, 49, 60, 63
absorption spectroscopy, vii, 42, 55, 65
accuracy, 1, 20, 21
active site, 29, 30
Ag, 63
air, 31
algorithm, 26
alkali, 57
amorphous, 1, 2, 6, 42
amorphous phases, 42
amplitude, 3, 4, 12, 13, 17, 21, 42, 44, 47, 49, 54, 61
angular momentum, 10, 23, 24, 47
application, 25
argon, 7
assignment, 10
assumptions, 19, 29
atmosphere, 7
attribution, 1, 10, 11, 12, 13, 15, 16, 17, 18, 50, 53, 63
averaging, 17

B

behavior, 55
black, 13
Boltzmann constant, 57
bonding, 29
bonds, 24, 27, 30, 31, 39
borate, 46
broad spectrum, 42

C

candidates, 47
capacity, 67
cell, 24, 27, 30, 31, 46
ceramic, 3, 4, 41, 42, 43
ceramics, vii, 6, 8, 65, 67
channels, 35
clusters, 2, 5, 39, 65
communication, 72
compensation, 7, 9, 47
complementary, 15
complexity, 50
complications, 17
components, 49
compounds, 29, 46, 47
computer, 7, 8
concentration, 2, 8, 9, 10, 12, 18, 35, 36, 37, 38, 39, 42, 43, 46, 58, 61

Index

configuration, 10, 26, 27
Congress, iv
contamination, 47
controlled, 7, 8
coordination, 27, 29
correlation, 30
Coulomb, 23, 26, 28, 32
Coulomb interaction, 23
coupling, 6, 24, 31, 47, 53, 55, 56, 57
coupling constants, 6, 53
covalent, 29
covalent bond, 29
covalent bonding, 29
covering, vii, 10, 20
CRC, 69
crystal, vii, 2, 3, 4, 5, 10, 11, 16, 20, 22, 25, 26, 27, 28, 29, 30, 31, 32, 33, 37, 39, 40, 42, 43, 46, 47, 48, 57, 58, 63, 67
crystal lattice, 10
crystal structure, 33
crystalline, 1, 2, 5, 33, 35, 39, 41, 53, 59, 61
crystallites, 41
crystals, vii, 1, 5, 7, 8, 9, 14, 15, 19, 20, 35, 37, 39, 48, 51, 53, 54, 55, 56, 62, 67

D

data base, 10
Debye, 55, 56, 57, 58
decay, 35, 55
decay times, 55
decoupling, 42
degenerate, 25, 48
detection, 51
deviation, 17
diffraction, 42, 46, 63
dipole, 21, 22, 28, 50, 60
disorder, 35, 37, 39
disordered systems, 54
distribution, 37, 41, 63
dopant, 9, 12, 19, 31, 32, 37, 39, 43, 54
dopants, 8, 30, 63
doped, vii, 1, 2, 5, 7, 11, 14, 17, 19, 20, 27, 31, 36, 39, 41, 42, 43, 45, 46, 47, 53, 54, 57, 58, 60, 61, 62

doping, 8, 9, 37, 39, 45, 46, 58, 65
drying, 7

E

earth, vii, 25, 26, 27, 28, 32, 41, 46, 47
eigenvalues, 26
election, 21
electric field, 9, 15, 16, 21, 25, 44
electron, 1, 2, 6, 23, 25, 29, 42, 55, 65
electron diffraction, 42
electron microscopy, 41
Electron Paramagnetic Resonance, 72
electronic, iv, vii, 6, 30, 47, 50, 51, 53, 55, 59, 61, 63
electronic structure, 30
electron-phonon, 2, 6, 55
electrons, 10, 15, 16, 23, 25, 26, 35, 37, 46, 51, 53, 55, 56
electrostatic, iv, 1, 10, 28, 32
emission, vii, 35, 41, 53, 57
energy, vii, 6, 10, 13, 14, 15, 19, 20, 25, 26, 27, 30, 31, 33, 35, 41, 47, 53, 55, 57, 58, 61, 62, 65
energy level splitting, 26
envelope, 39
environment, vii, 2, 25, 42, 43, 56
erbium, 1, 7, 33
estimating, 21
evidence, 5, 24, 27, 30, 31, 38, 41, 46, 61, 63
excitation, 6, 55
experimental condition, 7
expert, iv
exponential, 39
eyes, 13, 16, 38, 45

F

fiber, 1
field theory, vii, 5, 28
films, 7
financial support, 67
fine tuning, 45
fluorescence, 20, 21, 55

fluorides, 33
fluorinated, 9, 39, 40, 42, 43, 53
fluorine, 33, 39
Fourier, 9, 70

G

Gaussian, 37, 54
gel, 8, 9, 39, 40
gelation, 7
glass, vii, 3, 4, 6, 8, 39, 41, 42, 43, 65, 67
glasses, vii, 7, 8, 9, 41, 43, 54, 67
gold, 9
graduate students, 67
grain, 42
grain boundaries, 42
groups, 8, 39, 46, 49, 51
growth, 7, 8, 18
guidelines, 61

H

H1, 22
halogen, 9
Hamiltonian, 25, 26, 27, 28, 31, 51
harmonics, 25
heat, 57, 67
heat capacity, 67
high pressure, 9
high resolution, 1, 2, 5, 10, 15, 17, 20, 21, 23, 35, 42, 43, 47, 51, 57, 65, 67
high temperature, 21, 54
histogram, 62, 63
homogeneous, 36, 39, 43, 49, 54
host, 1, 2, 5, 12, 27, 29, 35, 37, 44, 57, 58, 63
Hungary, 7, 67
hyperfine interaction, vii, 47, 50
hypothesis, 25, 45, 47

I

identification, 1, 13, 18, 21
impurities, 37
inclusion, 14, 42

infrared, 9, 47
injury, iv
insight, 23, 53, 65
inspection, 2, 58
integrated optics, vii, 5
integration, 58
intensity, 15, 17, 37, 39
interaction, iii, v, vii, 1, 2, 6, 10, 23, 25, 26, 27, 36, 44, 45, 46, 47, 50, 53, 55, 58, 60, 72
interactions, vii, 2, 5, 23, 26, 27, 37, 47, 51, 65
interface, 42
intermetallic compounds, 29
interpretation, vii, 5, 10, 29, 50
intrinsic, 29
ionic, 56
ions, 9, 10, 11, 16, 17, 25, 28, 29, 32, 33, 39, 41, 42, 45, 64
IR, 9, 10, 19, 58, 59, 62, 63
IR spectra, 62
isotopes, 37
Italy, 7, 70

K

KBr, 8, 9, 14

L

L1, 22
lanthanide, 5, 10, 33, 56
laser, vii, 1, 5, 41, 53
lattice, vii, 6, 36, 53, 61, 62, 63, 65
law, 29
lifetime, 35, 37, 55
ligand, 29, 31, 33
ligands, 29
linear, 15, 22, 23, 41
links, 29
liquid nitrogen, 9
literature, 65
localization, 1
London, 70
low temperatures, vii, 1, 10, 11, 55, 59

Index

luminescence, 39
lying, 13, 15, 35, 55, 57

M

M1, 22
magnetic, iv, vii, 21, 22, 25, 26, 27, 28, 30, 48, 50
magnetic field, 25, 48
magnetic moment, vii
main line, 2, 36, 44
manifold, 2, 10, 11, 12, 13, 14, 15, 16, 17, 20, 21, 22, 24, 25, 28, 31, 33, 35, 44, 45, 47, 48, 51, 61, 63
manifolds, 2, 10, 11, 14, 15, 16, 17, 20, 24, 28, 33, 35, 44, 45
manipulation, 47
matrix, 1, 2, 5, 6, 17, 25, 26, 35, 37, 43, 58, 60, 63
mechanical, iv
media, 47
microscopy, 41
MIT, 71
modeling, 67
models, 6, 29, 53, 57, 58
molecules, 39
momentum, 10, 23, 24, 47, 51

N

nanoclusters, 5, 7, 41
nanocrystals, 2, 42, 53, 65
nanostructured materials, 1
natural, 37, 48
Nd, 30, 55
network, 39, 42
New York, iii, iv, 69, 71
nitrate, 7
nitrogen, 9
noise, 63
non-linear, 12, 41
non-linearity, 12
nuclear, vii, 47, 48, 51, 65
nuclear spins, 65

nucleus, 23, 25

O

on-line, 41
operator, 60
optical, vii, 1, 5, 7, 8, 11, 14, 19, 20, 26, 38, 39, 41, 44, 49, 59, 62, 65
optical properties, 26, 41
optics, vii, 5
orbit, 10, 23, 24, 26, 28, 31, 32
orthorhombic, 33
oscillator, 38, 39, 43
oxide, 7
oxides, 47
oxygen, 7, 9, 27, 47

P

paramagnetic, 29
parameter, 29, 31, 56, 57, 58
particles, 41
perturbation, 25, 47
phonon, iii, v, 1, 6, 18, 41, 48, 53, 55, 56, 57, 58, 59, 61, 62, 63
phonons, 6, 53, 55, 58, 59, 61, 62, 63, 64
photon, 53, 57, 61
play, 23, 40
polarization, 5, 16, 22, 44
polarized, 9, 16, 21, 23, 28
polarized light, 9, 21, 28
polycrystalline, 1, 2, 3, 4, 5, 7, 8, 14, 15, 42
polyethylene, 9
population, 12
powder, 40
powders, 3, 4, 5, 7, 8, 9, 14, 55, 63
power, 29
preparation, iv, 8
pressure, 9
private, 72
probability, 39
probe, 1, 2, 39, 42
property, iv

Q

QDs, 41
quanta, 63
quantum, 24, 29, 47, 60
quantum state, 29
quartz, 9

R

radiation, 9
radius, 25, 30, 56
Raman, 55, 56, 58, 59, 62, 63, 67
Raman scattering, 55, 56, 62
Raman scattering measurements, 62
Raman spectra, 59, 63, 67
Raman spectroscopy, 58
random, 37, 63
range, vii, 2, 5, 6, 9, 10, 11, 17, 18, 19, 20, 28, 30, 32, 33, 36, 37, 46, 51, 57, 63, 65
rare earth, vii, 5, 23, 27, 30, 31, 41, 48
rare earths, vii, 5, 23, 27, 30, 31, 48
recalling, 26
red shift, 54
reference frame, 30
relaxation, 55
resolution, vii, 1, 2, 5, 9, 10, 15, 17, 20, 21, 23, 35, 38, 42, 43, 47, 48, 49, 51, 55, 57, 59, 65, 67
room temperature, 1, 3

S

sample, 2, 3, 4, 9, 10, 11, 13, 16, 18, 20, 22, 36, 39, 42, 43, 44, 45, 46, 48, 49, 56, 58, 60, 61, 62, 63
scattering, 55, 56, 62
scintillators, 1
segregation, 7, 41
sensitivity, 46
separation, 13, 15, 16, 17, 20, 21
series, 5, 6, 10, 19, 28, 30, 45, 56
services, iv
shape, 6, 37, 54
shoulder, 17
silica, 2, 6, 7, 37, 39, 40, 41, 42, 53
silicon, 9
single crystals, vii, 5, 7, 8, 9, 14, 15, 19, 20, 55
SiO_2, 3, 4, 8, 37, 39, 40, 41, 42, 43
sites, 37, 42, 44, 45, 51
sol-gel, 7, 8, 9, 39, 40
solid solutions, 37, 59
solubility, 7
solutions, 37, 59
specific heat, 57
spectra, vii, 1, 2, 3, 4, 5, 9, 10, 11, 13, 14, 15, 16, 17, 18, 19, 20, 21, 23, 28, 31, 33, 35, 36, 37, 38, 39, 40, 41, 42, 43, 44, 47, 48, 49, 50, 53, 54, 55, 56, 59, 60, 61, 62, 63, 67
spectral analysis, 14
spectroscopy, vii, 5, 42, 47, 51, 55, 58, 65
spectrum, 1, 12, 13, 16, 17, 19, 21, 39, 42, 58, 61, 62, 63
spin, 10, 23, 24, 26, 28, 31, 32, 47, 48
standard deviation, 17
storage, 1
strain, 37
strength, 38, 39
stretching, 37, 57
students, 67
substitutes, 9, 40
substitution, 7, 9
Sun, 69
superposition, vii, 5, 30, 31, 33, 42, 46, 50, 56
supply, vii, 10, 45, 58
symmetry, 10, 16, 21, 23, 25, 26, 27, 28, 29, 30, 33, 37, 39, 40, 48, 50, 64
synthesis, 39
systematic, 27
systems, 21, 32, 41, 54, 64

T

technology, 1, 41
telecommunications, 41
TEM, 41

Index

temperature, vii, 1, 2, 3, 5, 6, 9, 10, 11, 12, 13, 15, 17, 18, 19, 20, 21, 35, 37, 39, 42, 43, 44, 47, 48, 49, 53, 54, 55, 56, 57, 58, 65
temperature dependence, 10, 12, 15, 17, 18, 19, 54, 55, 56, 57, 58
TEOS, 7
textbooks, 1
theoretical, 2, 10, 17, 65, 67
theory, vii, 5, 25, 28
thermal, 39, 41
thermal treatment, 39, 41
third order, 41
time, 21, 25, 48
tin, 7, 9, 41
TiO_2, 9
transition, 1, 2, 3, 4, 5, 6, 10, 11, 12, 13, 16, 17, 18, 19, 20, 21, 22, 35, 36, 38, 39, 40, 41, 42, 43, 44, 45, 49, 50, 53, 54, 56, 58, 59, 60, 61, 62, 63, 64
transitions, vii, 1, 2, 10, 11, 13, 14, 15, 16, 17, 18, 19, 20, 22, 23, 28, 35, 44, 46, 48, 49, 50, 51, 53, 55, 60, 61, 63, 65
transmission, 41
transmission electron microscopy, 41
transparency, 14
transparent, 1, 9, 41
trend, 1, 21, 23, 35, 39, 54, 56, 59
tunneling, 47

U

ultraviolet, 10
uncertainty, 46
undergraduate, 67

V

vacuum, 9
validity, 26, 28, 29
values, 10, 14, 19, 23, 27, 28, 29, 30, 31, 32, 51, 55, 56, 58, 62, 63
variance, 9, 18, 43, 57, 58
vibrational, 6, 47, 58, 59, 62, 63
vibrational modes, 58, 60, 62, 63
visible, 9, 33

W

wave number, 5, 9, 10, 14, 17, 18, 20, 35, 36, 44, 51, 54, 57, 60, 61, 62, 63
weak interaction, 5
weakness, 45, 54
weight ratio, 8, 9
windows, 9
workers, 67

X

x-ray, 7, 45, 63
x-ray diffraction, 45, 63

Y

yttrium, 46

Z

zero-phonon lines, 48